Data and Fractions

Three out of Four Like Spaghetti

Grade 4

Also appropriate for Grade 5

Mary Berle-Carman
Karen Economopoulos
Andee Rubin
Susan Jo Russell

Contributing Author
Rebecca B. Corwin

Developed at TERC, Cambridge, Massachusetts

Dale Seymour Publications®
Menlo Park, California

The *Investigations* curriculum was developed at TERC (formerly
Technical Education Research Centers) in collaboration with Kent State
University and the State University of New York at Buffalo. The work was
supported in part by National Science Foundation Grant No. ESI-9050210.
TERC is a nonprofit company working to improve mathematics and science
education. TERC is located at 2067 Massachusetts Avenue, Cambridge,
MA 02140.

This project was supported, in part,
by the
National Science Foundation
Opinions expressed are those of the authors
and not necessarily those of the Foundation

Managing Editor: Catherine Anderson
Series Editor: Beverly Cory
Revision Team: Laura Marshall Alavosus, Ellen Harding, Patty Green Holubar,
Suzanne Knott, John Lanyi, Beverly Hersh Lozoff
ESL Consultant: Nancy Sokol Green
Production/Manufacturing Director: Janet Yearian
Production/Manufacturing Coordinator: Amy Changar, Shannon Miller
Design Manager: Jeff Kelly
Design: Don Taka
Illustrations: Rhonda Henrichsen, Barbara Epstein-Eagle, Rebecca Krug
Cover: Bay Graphics
Composition: Archetype Book Composition

This book is published by Dale Seymour Publications®, an imprint of
Addison Wesley Longman, Inc.

Dale Seymour Publications
2725 Sand Hill Road
Menlo Park, CA 94025
Customer Service: 800-872-1100

Order number DS47000
ISBN 1-57232-753-7
1 2 3 4 5 6 7 8 9 10-ML-01 00 99 98 97

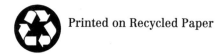
Printed on Recycled Paper

INVESTIGATIONS IN NUMBER, DATA, AND SPACE®

T E R C

Principal Investigator   Susan Jo Russell

Co-Principal Investigator   Cornelia C. Tierney

Director of Research and Evaluation   Jan Mokros

Curriculum Development
Joan Akers
Michael T. Battista
Mary Berle-Carman
Douglas H. Clements
Karen Economopoulos
Ricardo Nemirovsky
Andee Rubin
Susan Jo Russell
Cornelia C. Tierney
Amy Shulman Weinberg

Evaluation and Assessment
Mary Berle-Carman
Abouali Farmanfarmaian
Jan Mokros
Mark Ogonowski
Amy Shulman Weinberg
Tracey Wright
Lisa Yaffee

Teacher Support
Rebecca B. Corwin
Karen Economopoulos
Tracey Wright
Lisa Yaffee

Technology Development
Michael T. Battista
Douglas H. Clements
Julie Sarama Meredith
Andee Rubin

Video Production
David A. Smith

Cooperating Classrooms for This Unit
Marquita Jackson
Boston Public Schools
Boston, MA

Sarah Napier
Isabel Eccles
Fayerweather Street School
Cambridge, MA

Helen McElroy
Cambridge Friends School
Cambridge, MA

Consultants and Advisors
Elizabeth Badger
Deborah Lowenberg Ball
Marilyn Burns
Ann Grady
Joanne M. Gurry
James J. Kaput
Steven Leinwand
Mary M. Lindquist
David S. Moore
John Olive
Leslie P. Steffe
Peter Sullivan
Grayson Wheatley
Virginia Woolley
Anne Zarinnia

Administration and Production
Amy Catlin
Amy Taber

Graduate Assistants
Kent State University
Joanne Caniglia
Pam DeLong
Carol King

State University of New York at Buffalo
Rosa Gonzalez
Sue McMillen
Julie Sarama Meredith
Sudha Swaminathan

Revisions and Home Materials
Cathy Miles Grant
Marlene Kliman
Margaret McGaffigan
Megan Murray
Kim O'Neil
Andee Rubin
Susan Jo Russell
Lisa Seyferth
Myriam Steinback
Judy Storeygard
Anna Suarez
Cornelia Tierney
Carol Walker
Tracey Wright

CONTENTS

WHERE TO START

The first-time user of *Three out of Four Like Spaghetti* should read the following:

When you next teach this same unit, you can begin to read more of the background. Each time you present the unit, you will learn more about how your students understand the mathematical ideas.

Investigations in Number, Data, and Space® is a K–5 mathematics curriculum with four major goals:

- to offer students meaningful mathematical problems
- to emphasize depth in mathematical thinking rather than superficial exposure to a series of fragmented topics
- to communicate mathematics content and pedagogy to teachers
- to substantially expand the pool of mathematically literate students

The *Investigations* curriculum embodies a new approach based on years of research about how children learn mathematics. Each grade level consists of a set of separate units, each offering 2–8 weeks of work. These units of study are presented through investigations that involve students in the exploration of major mathematical ideas.

Approaching the mathematics content through investigations helps students develop flexibility and confidence in approaching problems, fluency in using mathematical skills and tools to solve problems, and proficiency in evaluating their solutions. Students also build a repertoire of ways to communicate about their mathematical thinking, while their enjoyment and appreciation of mathematics grows.

The investigations are carefully designed to invite all students into mathematics—girls and boys, members of diverse cultural, ethnic, and language groups, and students with different strengths and interests. Problem contexts often call on students to share experiences from their family, culture, or community. The curriculum eliminates barriers—such as work in isolation from peers, or emphasis on speed and memorization—that exclude some students from participating successfully in mathematics. The following aspects of the curriculum ensure that all students are included in significant mathematics learning:

- Students spend time exploring problems in depth.
- They find more than one solution to many of the problems they work on.

- They invent their own strategies and approaches, rather than relying on memorized procedures.
- They choose from a variety of concrete materials and appropriate technology, including calculators, as a natural part of their everyday mathematical work.
- They express their mathematical thinking through drawing, writing, and talking.
- They work in a variety of groupings—as a whole class, individually, in pairs, and in small groups.
- They move around the classroom as they explore the mathematics in their environment and talk with their peers.

While reading and other language activities are typically given a great deal of time and emphasis in elementary classrooms, mathematics often does not get the time it needs. If students are to experience mathematics in depth, they must have enough time to become engaged in real mathematical problems. We believe that a minimum of five hours of mathematics classroom time a week—about an hour a day—is critical at the elementary level. The plan and pacing of the *Investigations* curriculum is based on that belief.

We explain more about the pedagogy and principles that underlie these investigations in Teacher Notes throughout the units. For correlations of the curriculum to the NCTM Standards and further help in using this research-based program for teaching mathematics, see the following books:

- *Implementing the* Investigations in Number, Data, and Space® *Curriculum*
- *Beyond Arithmetic: Changing Mathematics in the Elementary Classroom* by Jan Mokros, Susan Jo Russell, and Karen Economopoulos

This book is one of the curriculum units for *Investigations in Number, Data, and Space.* In addition to providing part of a complete mathematics curriculum for your students, this unit offers information to support your own professional development. You, the teacher, are the person who will make this curriculum come alive in the classroom; the book for each unit is your main support system.

Although the curriculum does not include student textbooks, reproducible sheets for student work are provided in the unit and are also available as Student Activity Booklets. Students work actively with objects and experiences in their own environment and with a variety of manipulative materials and technology, rather than with a book of instruction and problems. We strongly recommend use of the overhead projector as a way to present problems, to focus group discussion, and to help students share ideas and strategies.

Ultimately, every teacher will use these investigations in ways that make sense for his or her particular style, the particular group of students, and the constraints and supports of a particular school environment. Each unit offers information and guidance for a wide variety of situations, drawn from our collaborations with many teachers and students over many years. Our goal in this book is to help you, a professional educator, implement this curriculum in a way that will give all your students access to mathematical power.

Investigation Format

The opening two pages of each investigation help you get ready for the work that follows.

What Happens This gives a synopsis of each session or block of sessions.

Mathematical Emphasis This lists the most important ideas and processes students will encounter in this investigation.

What to Plan Ahead of Time These lists alert you to materials to gather, sheets to duplicate, transparencies to make, and anything else you need to do before starting.

Using Fractions to Describe Data

What Happens

Session 1: Playing Guess My Rule Students play Guess My Rule as a way of introducing the fraction language and notation used throughout this unit—for example, "14 out of 26, or 14⁄26, of the students in the class are wearing long pants."

Session 2: Finding Familiar Fractions Students review the fractions ¼, ⅓, ½, ⅔, and ¾ by folding strips of paper into fractional parts. They use Class Strips to illustrate data about a group of people. By folding the Class Strips into halves, quarters, or thirds, they are able to describe such fractions as 14⁄26 in more familiar terms—for example, "About ¾ of us like spaghetti."

Session 3: Comparing Data with Familiar Fractions Students use familiar fractions to represent data about themselves and their families and to compare themselves with the country as a whole.

Session 4: Using Fractions to Compare Data Students discuss and compare their class data to the national data using their work from the previous session. Students do word problems that require them to relate fractions to real situations—for example, "How many people are three-quarters of the class?"

Mathematical Emphasis

- Partitioning a group according to a rule (for example, those who are wearing sweatshirts and those who are not)
- Finding familiar fractions (½, ¼, ⅓) of a group
- Estimating complex fractions with familiar fractions (for example, 12⁄25 is about ½)
- Collecting and analyzing categorical data (for example, nonnumerical responses to such questions as: "What is your favorite sport?")
- Describing data in terms of fractions (for example, "About ¾ of us like spaghetti")
- Using fractions to compare data from two groups, including two groups of different sizes
- Recognizing that fractions are always fractions of a particular whole

"ABOUT ¼ OF AMERICANS 8–17 HAVE 4 LIVING GRANDPARENTS."

What to Plan Ahead of Time

Materials

- Strip of paper to make class strip (Session 1, optional).
- Adding machine tape cut into 18-inch strips: 3–4 strips per student (Session 2)
- Markers or crayons, scissors, tape (Sessions 2–3)
- Interlocking cubes or counters: about 30 per student (Sessions 2–4)
- Overhead projector (Sessions 2–4)
- Clear container filled with two colors of beans, blocks, tiles, or some other material of similar size and shape (Ten-Minute Math)

Other Preparation

- Duplicate student sheets and teaching resources (located at the end of this unit) in the following quantities. If you have Student Activity Booklets, copy only the item marked with an asterisk, including any transparencies and extra materials needed.

  For Session 2
  Class Strips (p. 67): 5–6 sheets per student plus extras* (to be used throughout the unit), and 1 overhead transparency*
  Student Sheet 1, Finding Familiar Fractions (p. 57): 2 per student (1 for homework)
  Family letter* (p. 56): 1 per student. Remember to sign it before copying.

  For Session 3
  Student Sheet 2 (pages 1 and 2), Comparing Class Data and National Data (p. 58): 1 per student, and 1 overhead transparency*

  For Session 4
  Student Sheet 3, Word Problems (p. 60): 1 per student, homework

- Make a fraction strip from adding machine tape yourself before Session 2 in order to understand the task facing students. Try using Class Strips yourself to find familiar fractions of a few of the fractions from Guess My Rule.
- Make a large fraction strip showing the fractions ¼, ⅓, ½, ⅔, and ¾, just like the fraction strips students make in Session 2, but larger. You can make it by folding and labeling a 24-inch strip of adding machine tape or oak tag. Post this as a reference for the class after students make their own fraction strip in Session 2.
- If you plan to provide folders in which students will save their work for the entire unit, prepare these for distribution during Session 1.

Sessions Within an investigation, the activities are organized by class session, a session being at least a one-hour math class. Sessions are numbered consecutively through an investigation. Often several sessions are grouped together, presenting a block of activities with a single major focus.

When you find a block of sessions presented together—for example, Sessions 1, 2, and 3—read through the entire block first to understand the overall flow and sequence of the activities. Make some preliminary decisions about how you will divide the activities into three sessions for your class, based on what you know about your students. You may need to modify your initial plans as you progress through the activities, and you may want to make notes in the margins of the pages as reminders for the next time you use the unit.

Be sure to read the Session Follow-Up section at the end of the session block to see what homework assignments and extensions are suggested as you make your initial plans.

While you may be used to a curriculum that tells you exactly what each class session should cover, we have found that the teacher is in a better position to make these decisions. Each unit is flexible and may be handled somewhat differently by every teacher. While we provide guidance for how many sessions a particular group of activities is likely to need, we want you to be active in determining an appropriate pace and the best transition points for your class. It is not unusual for a teacher to spend more or less time than is proposed for the activities.

Ten-Minute Math At the beginning of some sessions, you will find Ten-Minute Math activities. These are designed to be used in tandem with the investigations, but not during the math hour. Rather, we hope you will do them whenever you have a spare 10 minutes—maybe before lunch or recess, or at the end of the day.

Ten-Minute Math offers practice in key concepts, but not always those being covered in the unit. For example, in a unit on using data, Ten-Minute Math might revisit geometric activities done earlier in the year. Complete directions for the suggested activities are included at the end of each unit.

Session 2

Finding Familiar Fractions

What Happens

Students review the fractions ¼, ⅓, ½, ⅔, and ¾ by folding strips of paper into fractional parts. They use Class Strips to illustrate data about a group of people. By folding the Class Strips into halves, quarters, or thirds, they are able to describe such fractions as 14⁄26 in more familiar terms—for example, "a little more than ½." Their work focuses on:

■ reviewing the size of familiar fractions relative to one another

■ understanding the size of unfamiliar fractions by identifying which familiar fractions they are close to

Materials
- Strips of adding machine tape 18 inches long (3–4 strips per student)
- Interlocking cubes or counters (30 per student)
- Markers or crayons, scissors, tape
- Class Strips (2–3 sheets per student)
- Transparency of Class Strips
- Student Sheet 1 (2 per student, homework)
- Large fraction strip prepared in advance
- Family letter (1 per student)
- Overhead projector

Activity

Making a Fraction Strip

Pass out one 18-inch strip of adding machine tape to each student. Have students fold the strip in half, mark ½ on the fold, and draw a line *lightly* on the fold as on a ruler so they can see it more clearly. Make sure students do not label the area between the folds. In order to see the relationships among the fractions clearly, students need to label the folds, not the areas between them.

$$\frac{1}{2}$$

When you folded your strips in half, you ended up with two equal parts. How could you fold the same strip of paper so you end up with four equal parts?

Some students will suggest that you fold the strip in half and then in half again. Other students might suggest you fold the strip in half, open it up, and then fold each half in half by folding the ends into the center. Either of these methods will yield four equal parts.

10 ■ *Investigation 1: Using Fractions to Describe Data*

Activities The activities include pair and small-group work, individual tasks, and whole-class discussions. In any case, students are seated together, talking and sharing ideas during all work times. Students most often work cooperatively, although each student may record work individually.

Choice Time In some units, some sessions are structured with activity choices. In these cases, students may work simultaneously on different activities focused on the same mathematical ideas. Students choose which activities they want to do, and they cycle through them.

You will need to decide how to set up and introduce these activities and how to let students make their choices. Some teachers present them as station activities, in different parts of the room. Some list the choices on the board as reminders or have students keep their own lists.

Extensions Sometimes in Session Follow-Up, you will find suggested extension activities. These are opportunities for some or all students to explore a topic in greater depth or in a different context. They are not designed for "fast" students; mathematics is a multifaceted discipline, and different students will want to go further in different investigations. Look for and encourage the sparks of interest and enthusiasm you see in your students, and use the extensions to help them pursue these interests.

Excursions Some of the *Investigations* units include excursions—blocks of activities that could be omitted without harming the integrity of the unit. This is one way of dealing with the great depth and variety of elementary mathematics— much more than a class has time to explore in any one year. Excursions give you the flexibility to make different choices from year to year, doing the excursion in one unit this time, and next year trying another excursion.

Tips for the Linguistically Diverse Classroom At strategic points in each unit, you will find concrete suggestions for simple modifications of the teaching strategies to encourage the participation of all students. Many of these tips offer alternative ways to elicit critical thinking from students at varying levels of English proficiency, as well as from other students who find it difficult to verbalize their thinking.

The tips are supported by suggestions for specific vocabulary work to help ensure that all students can participate fully in the investigations. The Preview for the Linguistically Diverse Classroom (p. I-18) lists important words that are assumed as part of the working vocabulary of the unit. Second-language learners will need to become familiar with these words in order to understand the problems and activities they will be doing. These terms can be incorporated into students' second-language work before or during the unit. Activities that can be used to present the words are found in the appendix, Vocabulary Support for Second-Language Learners (p. 53). In addition, ideas for making connections to students' language and cultures, included on the Preview page, help the class explore the unit's concepts from a multicultural perspective.

Materials

A complete list of the materials needed for teaching this unit is found on p. I-15. Some of these materials are available in kits for the *Investigations* curriculum. Individual items can also be purchased from school supply dealers.

Classroom Materials In an active mathematics classroom, certain basic materials should be available at all times: interlocking cubes, pencils, unlined paper, graph paper, calculators, things to count with, and measuring tools. Some activities in this curriculum require scissors and glue sticks or tape. Stick-on notes and large paper are also useful materials throughout.

So that students can independently get what they need at any time, they should know where these materials are kept, how they are stored, and how they are to be returned to the storage area. For example, interlocking cubes are best stored in towers of ten; then, whatever the activity, they should be returned to storage in groups of ten at the end of the hour. You'll find that establishing such routines at the beginning of the year is well worth the time and effort.

Technology Calculators are used throughout *Investigations*. Many of the units recommend that you have at least one calculator for each pair. You will find calculator activities, plus Teacher Notes discussing this important mathematical tool, in an early unit at each grade level. It is assumed that calculators will be readily available for student use.

Computer activities at grade 4 use a software program that was developed especially for the *Investigations* curriculum. The program *Geo-Logo*™ is used for activities in the 2-D Geometry unit, *Sunken Ships and Grid Patterns,* where students explore coordinate graphing systems, the use of negative numbers to represent locations in space, and the properties of geometric figures.

How you use the computer activities depends on the number of computers you have available. Suggestions are offered in the geometry units for how to organize different types of computer environments.

Children's Literature Each unit offers a list of suggested children's literature (p. I-15) that can be used to support the mathematical ideas in the unit. Sometimes an activity is based on a specific children's book, with suggestions for substitutions where practical. While such activities can be adapted and taught without the book, the literature offers a rich introduction and should be used whenever possible.

Student Sheets and Teaching Resources Student recording sheets and other teaching tools needed for both class and homework are provided as reproducible blackline masters at the end of each unit. They are also available as Student Activity Booklets. These booklets contain all the sheets each student will need for individual work, freeing you from extensive copying (although you may need or want to copy the occasional teaching resource on transparency film or card stock, or make extra copies of a student sheet).

We think it's important that students find their own ways of organizing and recording their work. They need to learn how to explain their thinking with both drawings and written words, and how to organize their results so someone else can understand them. For this reason, we deliberately do not

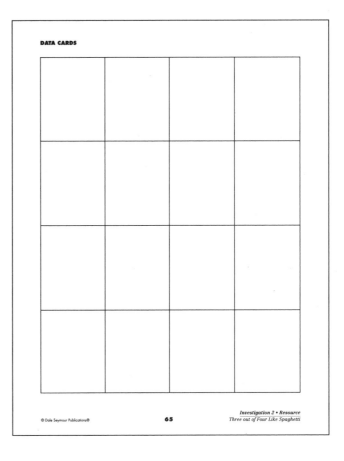

DATA CARDS

© Dale Seymour Publications®   65   *Investigation 2 • Resource*
Three out of Four Like Spaghetti

provide student sheets for every activity. Regardless of the form in which students do their work, we recommend that they keep a mathematics notebook or folder so that their work is always available for reference.

Homework In *Investigations,* homework is an extension of classroom work. Sometimes it offers review and practice of work done in class, sometimes preparation for upcoming activities, and sometimes numerical practice that revisits work in earlier units. Homework plays a role both in supporting students' learning and in helping inform families about the ways in which students in this curriculum work with mathematical ideas.

Depending on your school's homework policies and your own judgment, you may want to assign more homework than is suggested in the units. For this purpose you might use the practice pages, included as blackline masters at the end of this unit, to give students additional work with numbers.

For some homework assignments, you will want to adapt the activity to meet the needs of a variety of students in your class: those with special needs, those ready for more challenge, and second-language learners. You might change the numbers in a problem, make the activity more or less complex, or go through a sample activity with those who need extra help. You can modify any student sheet for either homework or class use. In particular, making numbers in a problem smaller or larger can make the same basic activity appropriate for a wider range of students.

Another issue to consider is how to handle the homework that students bring back to class—how to recognize the work they have done at home without spending too much time on it. Some teachers hold a short group discussion of different approaches to the assignment; others ask students to share and discuss their work with a neighbor, or post the homework around the room and give students time to tour it briefly. If you want to keep track of homework students bring in, be sure it ends up in a designated place.

Investigations at **Home** It is a good idea to make your policy on homework explicit to both students and their families when you begin teaching with *Investigations*. How frequently will you be assigning homework? When do you expect homework to be completed and brought back to school? What are your goals in assigning homework? How independent should families expect their children to be? What should the parent's or guardian's role be? The more explicit you can be about your expectations, the better the homework experience will be for everyone.

Investigations at Home (a booklet available separately for each unit, to send home with students) gives you a way to communicate with families about the work students are doing in class. This booklet includes a brief description of every session, a list of the mathematics content emphasized in each investigation, and a discussion of each homework assignment to help families more effectively support their children. Whether or not you are using the *Investigations* at Home booklets, we expect you to make your own choices about home-

work assignments. Feel free to omit any and to add extra ones you think are appropriate.

Family Letter A letter that you can send home to students' families is included with the black-line masters for each unit. Families need to be informed about the mathematics work in your classroom; they should be encouraged to participate in and support their children's work. A reminder to send home the letter for each unit appears in one of the early investigations. These letters are also available separately in Spanish, Vietnamese, Cantonese, Hmong, and Cambodian.

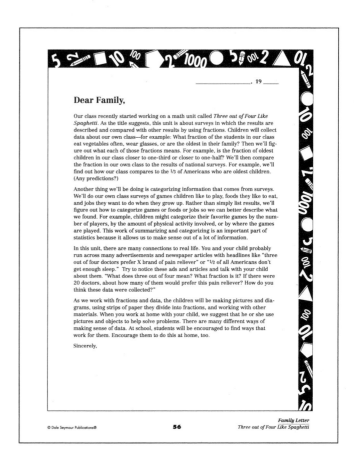

Help for You, the Teacher

Because we believe strongly that a new curriculum must help teachers think in new ways about mathematics and about their students' mathematical thinking processes, we have included a great deal of material to help you learn more about both.

About the Mathematics in This Unit This introductory section (p. I-16) summarizes the critical information about the mathematics you will be teaching. It describes the unit's central mathematical ideas and how students will encounter them through the unit's activities.

Teacher Notes These reference notes provide practical information about the mathematics you are teaching and about our experience with how students learn. Many of the notes were written in response to actual questions from teachers, or to discuss important things we saw happening in the field-test classrooms. Some teachers like to read them all before starting the unit, then review them as they come up in particular investigations.

Dialogue Boxes Sample dialogues demonstrate how students typically express their mathematical ideas, what issues and confusions arise in their thinking, and how some teachers have guided class discussions.

These dialogues are based on the extensive classroom testing of this curriculum; many are word-for-word transcriptions of recorded class discussions. They are not always easy reading; sometimes it may take some effort to unravel what the students are trying to say. But this is the value of these dialogues; they offer good clues to how your students may develop and express their approaches and strategies, helping you prepare for your own class discussions.

Where to Start You may not have time to read everything the first time you use this unit. As a first-time user, you will likely focus on understanding the activities and working them out with your students. Read completely through each investigation before starting to present it. Also read those sections listed in the Contents under the heading Where to Start (p. vi).

Teacher Note ***Comparing Groups of Different Sizes***

Students may ask how many people are in the national data set because they feel it is critical to know the number of people in each group in order to make a valid comparison. These students are probably thinking about comparing the *number* of people rather than comparing the *fraction* of people. Here are two ways to address the issue.

The population of the United States is so large that just about any fraction of it will be much larger than the number of students in your class. The population of the United States is about 250,000,000 (256,566,000 is the estimate from the U.S. Census Bureau for the population as of January 1993). So even one-fourth would be about 60 million. Comparing by number in this case just wouldn't make sense. We're not trying to see whether our class has more or fewer people who own dogs than in the whole country—we already know it's fewer! Rather, we're trying to see whether our class as a whole is similar to the country as a whole when we compare how big a part of our class owns dogs to how big a part of all Americans own dogs.

In Investigation 2, however, students compare their class with another class in their school, and the class sizes are likely to be close. In this case, some students may have an even harder time understanding why we use fractions rather than numbers to compare groups. The following example may be helpful. Suppose I collect data from two clubs about dogs. In one club, 50 out of 100 people own dogs. In another club, 5 out of 10 people own dogs.

Question	Club 1	Club 2
Own a dog?	$\frac{50}{100}$	$\frac{5}{10}$

How can we use these data to compare the two clubs?

Which club has a greater fraction of dog owners?

Don't be surprised if some of your students don't believe you can compare groups if there are unequal numbers in the groups. The idea that you can compare groups by using fractions (or percentages or averages), rather than directly comparing amounts, is a complex idea. It requires an understanding that a group can be considered as one whole and can be compared to another group also considered to be one whole. This idea requires that our view of the group change: We no longer consider size as an important attribute; we view each group as a unit. Much older students still have trouble with this idea, so don't expect all your students to understand it completely.

22 ▪ *Investigation 1: Using Fractions to Describe Data*

D I A L O G U E B O X

Finding the Familiar Fraction for 15/21

This discussion takes place during the activity Using Class Strips to Find Familiar Fractions (p. 12).

How could this strip help you figure out what 15/21 is close to? What would you say it's closest to: 1/2, 1/4, 2/4, 3/4, 1/3, or 2/3?

Alex: You could fold it in half. It would tell you that it's at least 1/2.

Nick: You could count off different fractions like 1/4 is 5. Fold the fraction that 15 is closest to, like if you thought it was fourths you could count off 1, 2, 3, 4, 5 sections and fold it, then 1, 2, 3, 4, 5 more and fold it—that's 1/2, then 5 more—that's 15. So 15 is closest to 3/4.

Luisa: I'd divide 21 into 3 parts, so that's 7. Then I'd count off 7 and fold the strip, count off 7 more and fold. Then it's a little more than 2/3.

Kim: If you fold it in half, then there are 10 on each side of the fold. I'm not counting the one in the middle. So it's more than 1/2.

Qi Sun: I think it's 2/3 because 1/3 of 21 is 7 and 2/3 of 21 is 14, so 15 is just 1 off.

Did you use a strip to help you think about the problem?

Qi Sun: [*picking up the strip and pointing to thirds*] If each of these sections represents 7, then 2 sections is 14, so I'd say that 15/21 is about 2/3.

Sarah: I did it the way Qi Sun did it.

Tell us how.

Sarah: Well, some people might say it's in quarters because 5, 10, 15, 20, and 20 is close to 21, but I think of it in thirds, and it's a little more than 2/3.

Luisa: First I thought it was going to be this one [*she points to quarters*], but I think it's this one in thirds because I made 21 into thirds and 2/3 is 14 out of 21.

Pinsuba: I thought that it's going to be thirds because 21 is an odd number and there are 3 parts in thirds and 4 parts in fourths, so thirds has an odd number, too.

Nick: I thought of it in fourths because each part is worth 5 and 15 is 3 parts, three quarters. If it was 20 people, then 15 out of 20 would be 3/4.

Qi Sun: But 1/4 of 21 is different than 1/4 of 20, so it's different.

16 ▪ *Investigation 1: Using Fractions to Describe Data*

The *Investigations* curriculum incorporates the use of two forms of technology in the classroom: calculators and computers. Calculators are assumed to be standard classroom materials, available for student use in any unit. Computers are explicitly linked to one or more units at each grade level; they are used with the unit on 2-D geometry at each grade, as well as with some of the units on measuring, data, and changes.

Using Calculators

In this curriculum, calculators are considered tools for doing mathematics, similar to pattern blocks or interlocking cubes. Just as with other tools, students must learn both *how* to use calculators correctly and *when* they are appropriate to use. This knowledge is crucial for daily life, as calculators are now a standard way of handling numerical operations, both at work and at home.

Using a calculator correctly is not a simple task; it depends on a good knowledge of the four operations and of the number system, so that students can select suitable calculations and also determine what a reasonable result would be. These skills are the basis of any work with numbers, whether or not a calculator is involved.

Unfortunately, calculators are often seen as tools to check computations with, as if other methods are somehow more fallible. Students need to understand that any computational method can be used to check any other; it's just as easy to make a mistake on the calculator as it is to make a mistake on paper or with mental arithmetic. Throughout this curriculum, we encourage students to solve computation problems in more than one way in order to double-check their accuracy. We present mental arithmetic, paper-and-pencil computation, and calculators as three possible approaches.

In this curriculum we also recognize that, despite their importance, calculators are not always appropriate in mathematics instruction. Like any tools, calculators are useful for some tasks, but

not for others. You will need to make decisions about when to allow students access to calculators and when to ask that they solve problems without them, so that they can concentrate on other tools and skills. At times when calculators are or are not appropriate for a particular activity, we make specific recommendations. Help your students develop their own sense of which problems they can tackle with their own reasoning and which ones might be better solved with a combination of their own reasoning and the calculator.

Managing calculators in your classroom so that they are a tool, and not a distraction, requires some planning. When calculators are first introduced, students often want to use them for everything, even problems that can be solved quite simply by other methods. However, once the novelty wears off, students are just as interested in developing their own strategies, especially when these strategies are emphasized and valued in the classroom. Over time, students will come to recognize the ease and value of solving problems mentally, with paper and pencil, or with manipulatives, while also understanding the power of the calculator to facilitate work with larger numbers.

Experience shows that if calculators are available only occasionally, students become excited and distracted when they are permitted to use them. They focus on the tool rather than on the mathematics. In order to learn when calculators are appropriate and when they are not, students must have easy access to them and use them routinely in their work.

If you have a calculator for each student, and if you think your students can accept the responsibility, you might allow them to keep their calculators with the rest of their individual materials, at least for the first few weeks of school. Alternatively, you might store them in boxes on a shelf, number each calculator, and assign a corresponding number to each student. This system can give students a sense of ownership while also helping you keep track of the calculators.

Using Computers

Students can use computers to approach and visualize mathematical situations in new ways. The computer allows students to construct and manipulate geometric shapes, see objects move according to rules they specify, and turn, flip, and repeat a pattern.

This curriculum calls for computers in units where they are a particularly effective tool for learning mathematics content. One unit on 2-D geometry at each of the grades 3–5 includes a core of activities that rely on access to computers, either in the classroom or in a lab. Other units on geometry, measurement, data, and changes include computer activities, but can be taught without them. In these units, however, students' experience is greatly enhanced by computer use.

The following list outlines the recommended use of computers in this curriculum:

Grade 1
Unit: *Survey Questions and Secret Rules* (Collecting and Sorting Data)
Software: Tabletop, Jr.
Source: Broderbund

Unit: *Quilt Squares and Block Towns* (2-D and 3-D Geometry)
Software: *Shapes*
Source: provided with the unit

Grade 2
Unit: *Mathematical Thinking at Grade 2* (Introduction)
Software: *Shapes*
Source: provided with the unit

Unit: *Shapes, Halves, and Symmetry* (Geometry and Fractions)
Software: *Shapes*
Source: provided with the unit

Unit: *How Long? How Far?* (Measuring)
Software: *Geo-Logo*
Source: provided with the unit

Grade 3
Unit: *Flips, Turns, and Area* (2-D Geometry)
Software: *Tumbling Tetrominoes*
Source: provided with the unit

Unit: *Turtle Paths* (2-D Geometry)
Software: *Geo-Logo*
Source: provided with the unit

Grade 4
Unit: *Sunken Ships and Grid Patterns* (2-D Geometry)
Software: *Geo-Logo*
Source: provided with the unit

Grade 5
Unit: *Picturing Polygons* (2-D Geometry)
Software: *Geo-Logo*
Source: provided with the unit

Unit: *Patterns of Change* (Tables and Graphs)
Software: *Trips*
Source: provided with the unit

Unit: *Data: Kids, Cats, and Ads* (Statistics)
Software: Tabletop, Sr.
Source: Broderbund

The software provided with the *Investigations* units uses the power of the computer to help students explore mathematical ideas and relationships that cannot be explored in the same way with physical materials. With the *Shapes* (grades 1–2) and *Tumbling Tetrominoes* (grade 3) software, students explore symmetry, pattern, rotation and reflection, area, and characteristics of 2-D shapes. With the *Geo-Logo* software (grades 3–5), students investigate rotations and reflections, coordinate geometry, the properties of 2-D shapes, and angles. The *Trips* software (grade 5) is a mathematical exploration of motion in which students run experiments and interpret data presented in graphs and tables.

We suggest that students work in pairs on the computer; this not only maximizes computer resources but also encourages students to consult, monitor, and teach one another. Generally, more than two students at one computer find it difficult to share. Managing access to computers is an issue for every classroom. The curriculum gives you explicit support for setting up a system. The units are structured on the assumption that you have enough computers for half your students to work on the machines in pairs at one time. If you do not have access to that many computers, suggestions are made for structuring class time to use the unit with five to eight computers, or even with fewer than five.

Assessment plays a critical role in teaching and learning, and it is an integral part of the *Investigations* curriculum. For a teacher using these units, assessment is an ongoing process. You observe students' discussions and explanations of their strategies on a daily basis and examine their work as it evolves. While students are busy recording and representing their work, working on projects, sharing with partners, and playing mathematical games, you have many opportunities to observe their mathematical thinking. What you learn through observation guides your decisions about how to proceed. In any of the units, you will repeatedly consider questions like these:

- Do students come up with their own strategies for solving problems, or do they expect others to tell them what to do? What do their strategies reveal about their mathematical understanding?

- Do students understand that there are different strategies for solving problems? Do they articulate their strategies and try to understand other students' strategies?

- How effectively do students use materials as tools to help with their mathematical work?

- Do students have effective ideas for keeping track of and recording their work? Does keeping track of and recording their work seem difficult for them?

You will need to develop a comfortable and efficient system for recording and keeping track of your observations. Some teachers keep a clipboard handy and jot notes on a class list or on adhesive labels that are later transferred to student files. Others keep loose-leaf notebooks with a page for each student and make weekly notes about what they have observed in class.

Assessment Tools in the Unit

With the activities in each unit, you will find questions to guide your thinking while observing the students at work. You will also find two built-in assessment tools: Teacher Checkpoints and embedded Assessment activities.

Teacher Checkpoints The designated Teacher Checkpoints in each unit offer a time to "check in" with individual students, watch them at work, and ask questions that illuminate how they are thinking.

At first it may be hard to know what to look for, hard to know what kinds of questions to ask. Students may be reluctant to talk; they may not be accustomed to having the teacher ask them about their work, or they may not know how to explain their thinking. Two important ingredients of this process are asking students open-ended questions about their work and showing genuine interest in how they are approaching the task. When students see that you are interested in their thinking and are counting on them to come up with their own ways of solving problems, they may surprise you with the depth of their understanding.

Teacher Checkpoints also give you the chance to pause in the teaching sequence and reflect on how your class is doing overall. Think about whether you need to adjust your pacing: Are most students fluent with strategies for solving a particular kind of problem? Are they just starting to formulate good strategies? Or are they still struggling with how to start? Depending on what you see as the students work, you may want to spend more time on similar problems, change some of the problems to use smaller numbers, move quickly to more challenging material, modify subsequent activities for some students, work on particular ideas with a small group, or pair students who have good strategies with those who are having more difficulty.

Embedded Assessment Activities Assessment activities embedded in each unit will help you examine specific pieces of student work, figure out what it means, and provide feedback. From the students' point of view, these assessment activities are no different from any others. Each is a learning experience in and of itself, as well as an opportunity for you to gather evidence about students' mathematical understanding.

The embedded assessment activities sometimes involve writing and reflecting; at other times, a discussion or brief interaction between student and teacher; and in still other instances, the creation and explanation of a product. In most cases, the assessments require that students *show* what they did, *write* or *talk* about it, or do both. Having to explain how they worked through a problem helps students be more focused and clear in their mathematical thinking. It also helps them realize that doing mathematics is a process that may involve tentative starts, revising one's approach, taking different paths, and working through ideas.

Teachers often find the hardest part of assessment to be interpreting their students' work. We provide guidelines to help with that interpretation. If you have used a process approach to teaching writing, the assessment in *Investigations* will seem familiar. For many of the assessment activities, a Teacher Note provides examples of student work and a commentary on what it indicates about student thinking.

Documentation of Student Growth

To form an overall picture of mathematical progress, it is important to document each student's work in journals, notebooks, or portfolios. The choice is largely a matter of personal preference; some teachers have students keep a notebook or folder for each unit, while others prefer one mathematics notebook, or a portfolio of selected work for the entire year. The final activity in each *Investigations* unit, called Choosing Student Work to Save, helps you and the students select representative samples for a record of their work.

This kind of regular documentation helps you synthesize information about each student as a mathematical learner. From different pieces of evidence, you can put together the big picture. This synthesis will be invaluable in thinking about where to go next with a particular child, deciding where more work is needed, or explaining to parents (or other teachers) how a child is doing.

If you use portfolios, you need to collect a good balance of work, yet avoid being swamped with an overwhelming amount of paper. Following are some tips for effective portfolios:

- Collect a representative sample of work, including some pieces that students themselves select for inclusion in the portfolio. There should be just a few pieces for each unit, showing different kinds of work—some assignments that involve writing, as well as some that do not.

- If students do not date their work, do so yourself so that you can reconstruct the order in which pieces were done.

- Include your reflections on the work. When you are looking back over the whole year, such comments are reminders of what seemed especially interesting about a particular piece; they can also be helpful to other teachers and to parents. Older students should be encouraged to write their own reflections about their work.

Assessment Overview

There are two places to turn for a preview of the assessment opportunities in each *Investigations* unit. The Assessment Resources column in the unit Overview Chart (pp. I-13–I-14) identifies the Teacher Checkpoints and Assessment activities embedded in each investigation, guidelines for observing the students that appear within classroom activities, and any Teacher Notes and Dialogue Boxes that explain what to look for and what types of student responses you might expect to see in your classroom. Additionally, the section About the Assessment in This Unit (p. I-17) gives you a detailed list of questions for each investigation, keyed to the mathematical emphases, to help you observe student growth.

Depending on your situation, you may want to provide additional assessment opportunities. Most of the investigations lend themselves to more frequent assessment, simply by having students do more writing and recording while they are working.

Three out of Four Like Spaghetti

Content of This Unit   Students collect, describe, display, and compare categorical data. Categorical data are non-numerical data—for example, responses to questions such as this: "What is your favorite spaghetti sauce?" Such data can be grouped into categories (for example, those who like red, white, or no sauce) and then can be analyzed using numbers, including fractions. Students play Guess My Rule, a game in which they try to discover a characteristic that is similar about a group of students in the class. Using data collected from this game, they use fractions to describe and compare parts of the class ("$9/28$ of the class wears watches," "$15/28$ owns dogs") and find familiar fractions, such as $1/2$ or $1/3$, that are more appropriate for reporting their data ("About $1/3$ of the class wear watches," "About $1/2$ own dogs"). In the second half of the unit, students focus on collecting and classifying data about games they like, foods they like, and careers that interest them. For each set of data, they try various ways of sorting the data to see what different classification schemes reveal about the data. In their final project, they collect data from their own class and from first graders in answer to the question, "What do you want to be when you grow up?" They classify, display, describe, and interpret the data to compare first and fourth graders' career plans.

Connections with Other Units   If you are doing the full-year *Investigations* curriculum in the suggested sequence for Grade 4, this is the eleventh of eleven units. In this unit students build on their understanding of fractions begun in the Fractions and Area unit, *Different Shapes, Equal Pieces*. This unit focuses on categorical data and complements the Statistics unit, *The Shape of the Data*, where students learned to represent and describe numerical data. In the Grade 5 fractions unit, students will use fraction strips for ordering fractions and finding equivalents.

This unit can be used successfully at either grade 4 or 5, depending on the prior experiences and needs of your students.

Investigations Curriculum   ■   Suggested Grade 4 Sequence

Mathematical Thinking at Grade 4 (Introduction)

Arrays and Shares (Multiplication and Division)

Seeing Solids and Silhouettes (3-D Geometry)

Landmarks in the Thousands (The Number System)

Different Shapes, Equal Pieces (Fractions and Area)

The Shape of the Data (Statistics)

Money, Miles, and Large Numbers (Addition and Subtraction)

Changes Over Time (Graphs)

Packages and Groups (Multiplication and Division)

Sunken Ships and Grid Patterns (2-D Geometry)

▶*Three out of Four Like Spaghetti* (Data and Fractions)

Investigation 1 ▪ Using Fractions to Describe Data

Class Sessions	Activities	Pacing
Session 1 (p. 4) PLAYING GUESS MY RULE	Guess My Rule Fractions to Describe Our Class Extension: Working with Two Rules at Once	minimum 1 hr
Session 2 (p. 10) FINDING FAMILIAR FRACTIONS	Making a Fraction Strip Estimating Familiar Fractions Using Class Strips to Find Familiar Fractions Homework: Finding Familiar Fractions	minimum 1 hr
Session 3 (p. 17) COMPARING DATA WITH FAMILIAR FRACTIONS	Teacher Checkpoint: More Data for Class Strips Comparing Our Class with National Data Homework: Comparing Class Data and National Data	minimum 1 hr
Session 4 (p. 23) USING FRACTIONS TO COMPARE DATA	Discussing Findings About the Data Using Fractions in Word Problems Homework: Word Problems	minimum 1 hr

◔ **Ten-Minute Math** ▪ **What Is Likely?**

Mathematical Emphasis

- Finding familiar fractions of a group

- Estimating complex fractions with familiar fractions

- Describing data in terms of fractions

- Using fractions to compare data from two groups, including two groups of different sizes

- Recognizing that fractions are always fractions of a particular whole

Assessment Resources

What Do You Notice About These Fractions? (Dialogue Box, p. 9)

Folding Familiar Fractions (Teacher Note, p. 15)

Finding the Familiar Fractions for 15/21 (Dialogue Box, p. 16)

Teacher Checkpoint: More Data for Class Strips (p. 17)

Problems with Wholes (Teacher Note, p. 21)

Comparing Groups of Different Sizes (Teacher Note, p. 22)

Materials

Overhead projector

Adding machine tape

Markers or crayons, scissors, tape

Interlocking cubes

Clear container filled with two colors of beans, blocks, tiles, or some other material of similar size and shape (Ten-Minute Math)

Family letter

Student Sheets 1–3

Teaching resource sheets

Investigation 2 ■ Looking at Data in Categories

Class Sessions	Activities	Pacing
Session 1 (p. 28) GAMES WE PLAY	What Games Do You Like? Homework: Games People Played Extension: Games from Different Cultures	minimum 1 hr
Session 2 (p. 32) MORE GAMES, AND WHAT HAVE WE EATEN?	Categorizing Favorite Games Teacher Checkpoint: Foods in the Last 24 Hours Homework: Transportation Data Extension: Games That Use Math	minimum 1 hr
Session 3 (p. 38) WHAT DO YOU WANT TO BE WHEN YOU GROW UP?	Collecting and Organizing Fourth Grade Data Preparing to Collect First Grade Data	minimum 1 hr
Session 4 (p. 42) ORGANIZING SOME FIRST AND FOURTH GRADE DATA	Combining First and Fourth Grade Data Sharing Categories in Small Groups	minimum 1 hr

Sessions 5, 6, and 7 (p. 45)
MAKING COMPARISONS WITH
ALL THE DATA
	Combining and Organizing the First and
	    Fourth Grade Data
	Final Projects: Graphing, Describing, and
	    Comparing Career Data
	Sharing Conclusions About the Career Data
	Assessment: Evaluating the Final Projects
	Choosing Student Work to Save
	Homework: What Do You Want to Be
	    When You Grow Up?
		minimum
		3 hr

◕ Ten-Minute Math ■ What Is Likely?

Mathematical Emphasis

- Collecting and recording categorical data

- Organizing data into categories, and making judgments about sets of categories

- Representing categorical data, including use of bar graphs

- Describing categorical data

- Using fractions to compare categorical data from two groups

Assessment Resources

Teacher Checkpoint: Foods in the Last 24 Hours (p. 34)

What Makes a Good Categorization? (Teacher Note, p. 36)

Making Categories (Dialogue Box, p. 41)

Sharing Categories (Dialogue Box, p. 44)

Assessment: Evaluating the Final Projects (p. 48)

Choosing Student Work to Save (p. 48)

Materials

Overhead projector (opt.)

Interlocking cubes, counters, and Class Strips from Investigation 1

Index cards and tape, or large stick-on notes

Chart paper

Crayons or markers, glue, scissors

Large paper (11-inch by 17-inch)

Paper clips

Envelopes

Blank paper for labels

Student Sheets 4–7

Teaching resource sheets

Following are the basic materials needed for the activities in this unit. Many of the items can be purchased from the publisher, either individually or in the Teacher Resource Package and the Student Materials Kit for grade 4. Detailed information is available on the *Investigations* order form. To obtain this form, call toll-free 1-800-872-1100 and ask for a Dale Seymour customer service representative.

Snap™ Cubes (interlocking cubes) or counters: about 30 per student

Adding machine tape cut into 18-inch strips: 3–4 strips per student

Large paper for making presentation graphs: 2 per pair (11-inch by 17-inch for Session 1 of Investigation 2; larger sheets for Sessions 5–7 of Investigation 2; you may need to tape several sheets together for each pair.)

Crayons or markers, scissors, glue, tape

Chart paper

Index cards and tape, or large stick-on notes: several per student, plus 10 extra

Paper clips: about 10 per pair of students

Envelopes: 1 per pair of students

Overhead projector

Clear container filled with two colors of beans, blocks, tiles, or some other material of similar size and shape (Ten-Minute Math)

The following materials are provided at the end of this unit as blackline masters. A Student Activity Booklet containing all student sheets and teaching resources needed for individual work is available.

Family Letter (p. 56)

Student Sheets 1–7 (p. 57)

Teaching Resources:

    Data Cards (p. 65)

    Making Comparisons with All the Career Data (p. 66)

    Class Strips (p. 67)

Practice Pages (p. 69)

Related Children's Literature

Reid, Margarette. *The Button Box*. New York: Dutton Children's Books, 1990.

This unit focuses on collecting, organizing, describing, and comparing data that can be placed into categories. Since fractions are a particularly useful tool in describing and comparing parts of groups, we highlight finding fractional parts of groups as a central element in students' work. Using fractions allows us to compare groups of different sizes. For example, if you want to compare the number of students in Grades K–2 who bring their lunches to school with the number of students in Grades 3–5 who bring their lunches to school, you might find out that about ¾ of the students in Grades 3–5 bring their lunches while only about ⅓ of those in grades K–2 bring their lunches. Classifying and then comparing data in this way often leads to important observations and further questions: Do young students like the school lunches better than older students? Do young students have less choice about whether they bring or buy their lunches? Do more of the older students make their own lunches or have more say in whether they bring or buy their lunches?

The unit begins with an investigation in which students describe data they collect about their own class by using fractions. An important part of this work is to find "familiar fractions"—such as ¼, ⅓, ½, ⅔, and ¾—that can be used in place of more unwieldy fractions that result from data collection. For example, if you took a survey of your class and found that 9 of your 28 students had not eaten breakfast that morning, you might report this to your colleagues by saying: "About ⅓ of the students in my class didn't eat breakfast this morning." Using the familiar fraction ⅓ in place of 9/28 makes the information more accessible to your audience. (In later grades, students learn to use percentages in the same ways they use fractions here.)

In the second half of the unit, students become more deeply involved in sorting into categories the data they collect and describing what they can see in their data once they have sorted them. Students have probably been very used to collecting categorical data during their work in the primary grades. Students often collect and graph data about favorites—favorite color, favorite ice cream flavor—or other categorical data, such as how they get to school. It can be difficult to get meaning from categorical data. The graphs are often flat and uninteresting. There is not much to say about this graph:

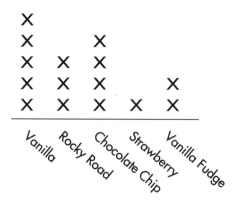

Typically, students describe data such as these by noticing which choice has the most, which has the least, and how many the others have: "Vanilla has the most," "Rocky Road has three," "Strawberry has the least."

There is something unsatisfying about this description. Is there anything of greater interest hidden in these data? In fact, grouping such data into categories often reveals other aspects of the story. In this case, even though vanilla seems to be the favorite in the graph, it looks like flavors that have some chocolate in them are really liked more:

Flavors with chocolate   //// ////

Flavors without chocolate   //// /

Classification is a critical process in science and mathematics. By examining similarities and differences to see what "goes together," we often reveal important information about the objects we are studying. As students find different ways to classify their data, they begin to see how different sets of categories can reveal different aspects of their data.

Mathematical Emphasis At the beginning of each investigation, the Mathematical Emphasis section tells you what is most important for students to learn about during that investigation. Many of these mathematical understandings and processes are difficult and complex. Students gradually learn more and more about each idea over many years of schooling. Individual students will begin and end the unit with different levels of knowledge and skill, but all will gain greater knowledge about classifying and describing categorical data and about using fractions to describe parts of groups.

Throughout the *Investigations* curriculum, there are many opportunities for ongoing daily assessment as you observe, listen to, and interact with students at work. In this unit, you will find two Teacher Checkpoints:

Investigation 1, Session 3:
More Data for Class Strips (p. 17)

Investigation 2, Session 2:
Foods in the Last 24 Hours (p. 34)

This unit also has one embedded assessment activity:

Investigation 2, Sessions 5, 6, and 7:
Evaluating the Final Projects (p. 48)

In addition, you can use almost any activity in this unit to assess your students' needs and strengths. Listed below are questions to help you focus your observations in each investigation. You may want to keep track of your observations for each student to help you plan your curriculum and monitor students' growth. Suggestions for documenting student growth can be found on page I-10, in the section About Assessment.

Investigation 1: Using Fractions to Describe Data

- Are students able to find familiar fractions (1/2, 1/4, 1/3) of a group?

- Are students able to estimate the size of unfamiliar fractions by identifying which familiar fractions they are close to (for example, 12/25 is about 1/2)? What strategies do the students use (for example, their knowledge of multiplication and division, folded strips, dealing cubes into piles)?

- Are students able to describe data in terms of fractions (for example, "About 3/4 of us like spaghetti")? How do the students make sense of the connection between pairs of fractions such as "10/29 of the students in our class are wearing sneakers, but 19/29 are not"? What meaning do they attach to the numerators and denominators of these fractions?

- How do students make use of fractions to compare data from two groups, including two groups of different sizes?

- How do students understand the relationship between the fraction and the whole? As students work with increasingly complex problems, do they maintain clarity about what the whole is and what fractional part they are considering?

Investigation 2: Looking at Data in Categories

- How do students go about collecting and recording categorical data?

- How flexibly do students think as they organize into categories the data they collect? Can they think of several different ways to group the data? Do their categories help them understand the data? Are students beginning to understand that they can get different information depending on how they categorize the data?

- How clearly do students' graphs depict the categories (for example, can you easily tell from their display which responses are from first graders and which are from fourth graders)? How do students use features such as color, patterns, location, or other methods to help the reader?

- When students use fractions to make comparisons, can they explain clearly why one fraction is larger than another? What kinds of models (familiar fractions? concrete models?) do students use to help them compare fractions?

In the *Investigations* curriculum, mathematical vocabulary is introduced naturally during the activities. We don't ask students to learn definitions of new terms; rather, they come to understand such words as *factor*, *area*, and *symmetry* by hearing them used frequently in discussion as they investigate new concepts. This approach is compatible with current theories of second-language acquisition, which emphasize the use of new vocabulary in meaningful contexts while students are actively involved with objects, pictures, and physical movement.

Listed below are some key words used in this unit that will not be new to most English speakers at this age level but may be unfamiliar to students with limited English proficiency. You will want to spend additional time working on these words with your students who are learning English. If your students are working with a second-language teacher, you might enlist your colleague's aid in familiarizing students with these words, before and during this unit. In the classroom, look for opportunities for students to hear and use these words. Activities you can use to present the words are given in the appendix, Vocabulary Support for Second-Language Learners (p. 53).

true, false During the first few sessions of the unit, students play a game called Guess My Rule. Students take turns trying to identify whether members of the class "fit" the hidden rule or not.

gameboard, board games, outdoor games In Session 1 of Investigation 2, students collect data about their favorite games, and categorize them. Categories may include board games, outdoor games, or any other distinctive aspects of the games.

foods, snacks, breakfast, dinner, lunch, this morning, yesterday, last night In Session 2 of Investigation 2, students collect data about some of the foods they eat over a 24-hour period. Students graph these data according to categories they make.

careers, jobs In Sessions 3 through 7 of Investigation 2, students collect data in their own class and from a first grade class, answering the question, "What do you want to be when you grow up?" They sort these career data into categories of their own making, compare the fraction of first graders and fourth graders within the categories, and report their findings.

Multicultural Extensions for All Students

Whenever possible, encourage students to share words, objects, customs, or any aspects of daily life from their own cultures and backgrounds that are relevant to the activities in this unit. For example:

- When students are gathering data about some of the foods they eat over a 24-hour period, use this as an opportunity to discuss foods from different cultures. If possible, invite parents or others from the community to help your students prepare some foods from different cultures.

- Encourage students to teach one another the games they play in their neighborhood or with their family.

Investigations

INVESTIGATION 1

Using Fractions to Describe Data

What Happens

Session 1: Playing Guess My Rule   Students play Guess My Rule as a way of introducing the fraction language and notation used throughout this unit—for example, "14 out of 26, or 14/26, of the students in the class are wearing long pants."

Session 2: Finding Familiar Fractions
Students review the fractions 1/4, 1/3, 1/2, 2/3, and 3/4 by folding strips of paper into fractional parts. They use Class Strips to illustrate data about a group of people. By folding the Class Strips into halves, quarters, or thirds, they are able to describe such fractions as 14/26 in more familiar terms—for example, "A little more than 1/2."

Session 3: Comparing Data with Familiar Fractions   Students use familiar fractions to represent data about themselves and their families and to compare themselves with the country as a whole.

Session 4: Using Fractions to Compare Data
Students discuss and compare their class data to the national data using their work from the previous session. Students do word problems that require them to relate fractions to real situations—for example, "How many people are three-quarters of the class?"

Mathematical Emphasis

- Partitioning a group according to a rule (for example, those who are wearing sweatshirts and those who are not)
- Finding familiar fractions (1/2, 1/4, 1/3) of a group
- Estimating complex fractions with familiar fractions (for example, 12/25 is about 1/2)
- Collecting and analyzing categorical data (for example, nonnumerical responses to such questions as: "What is your favorite sport?")
- Describing data in terms of fractions (for example, "About 3/4 of us like spaghetti")
- Using fractions to compare data from two groups, including two groups of different sizes
- Recognizing that fractions are always fractions of a particular whole

"ABOUT 1/4 OF AMERICANS 8-17 HAVE 4 LIVING GRANDPARENTS."

What to Plan Ahead of Time

Materials

- Strip of paper to make class strip (Session 1, optional)
- Adding machine tape cut into 18-inch strips: 3–4 strips per student (Session 2)
- Markers or crayons, scissors, tape (Sessions 2–3)
- Interlocking cubes or counters: about 30 per student (Sessions 2–4)
- Overhead projector (Sessions 2–4)
- Clear container filled with two colors of beans, blocks, tiles, or some other material of similar size and shape (Ten-Minute Math)

Other Preparation

- Duplicate student sheets and teaching resources (located at the end of this unit) in the following quantities. If you have Student Activity Booklets, copy only the item marked with an asterisk, including any transparencies and extra materials needed.

For Session 2

Class Strips (p. 67): 5–6 sheets per student plus extras* (to be used throughout the unit), and 1 overhead transparency*

Student Sheet 1, Finding Familiar Fractions (p. 57): 2 per student (1 for homework)

Family letter* (p. 56): 1 per student. Remember to sign it before copying.

For Session 3

Student Sheet 2 (pages 1 and 2), Comparing Class Data and National Data (p. 58): 1 per student, and 1 overhead transparency*

For Session 4

Student Sheet 3, Word Problems (p. 60): 1 per student, homework

- Make a fraction strip from adding machine tape yourself before Session 2 in order to understand the task facing students. Try using Class Strips yourself to find familiar fractions of a few of the fractions from Guess My Rule.
- Make a large fraction strip showing the fractions ¼, ⅓, ½, ⅔, and ¾, just like the fraction strips students make in Session 2, but larger. You can make it by folding and labeling a 24-inch strip of adding machine tape or oak tag. Post this as a reference for the class after students make their own fraction strip in Session 2.
- If you plan to provide folders in which students will save their work for the entire unit, prepare these for distribution during Session 1.

Playing Guess My Rule

Materials

■ Strip of paper to make class strip (optional)

What Happens

Students play Guess My Rule as a way of introducing the fraction language and notation used throughout this unit—for example, "14 people out of 26, or 14/26, of the students in the class are wearing long pants." Their work focuses on:

■ dividing the class into categories

■ identifying the fraction of the class in each category

Activity

Guess My Rule

Guess My Rule is a classification guessing game in which players try to figure out a characteristic that a set of objects has in common. In the version used here, the objects are the people in the classroom. The game engages students in logical thinking, using evidence, and testing theories; in this unit, we also use it to generate data about the class that can be described using fractions. A more extensive discussion of Guess My Rule is found in the **Teacher Note**, Playing Guess My Rule (p. 8).

Decide on a characteristic or rule that describes some of the class but not everyone. For at least the first game, choose a rule that is fairly obvious visually, such as "people wearing red." Start the game by telling students the names of a few people who fit the rule, then the names of a few people who don't. Those who fit the rule stand up on one side of the room and those who don't stand up on the other side. Then students take turns guessing which of the remaining students might or might not fit the rule, and you tell them whether they are correct. As more students who fit the rule are identified, have them stand up and join the group that fits the rule. Have students who don't fit the rule join that group.

Today we're going to play a game called Guess My Rule. I have in mind a mystery rule that describes something about the people in this class. This characteristic is something you can see. Some people in the class fit my rule and other people don't.

An example of a mystery rule is "everyone with brown eyes" (but that's not my rule). The object of the game is for you to guess my rule.

I'm going to start the game by telling you a few people who fit my rule. If I say you are someone who fits my rule, please stand next to the board. If I say you do not fit my rule, stand on the other side of the room. When you think you know what my rule is, don't tell me the rule. Instead, name someone still seated who you think fits the rule.

Allow students to guess the rule itself only after many students have had a chance to suggest someone who does or doesn't fit the rule. If no one guesses the rule, continue playing the game until every student is standing with one of the two groups. At this point many of the students might know the rule, but some may not. As students try to guess your rule, ask them to give reasons for their conjecture. It is possible students will come up with a rule that fits the evidence but is not what you had in mind. Acknowledge their good reasoning even though it did not lead to your rule.

❖ **Tip for the Linguistically Diverse Classroom** Have limited English proficient students "state" their rule by drawing a rebus picture and giving their reason by pointing to what each person has in common. For example, students draw glasses on the board (the rule), then point to the glasses of each person in the group (the reason).

Recording the Results After each game, record the number of students who fit the rule and compare them to the total number of students in the class.

For example, you might say:

There are 14 people in our class who are wearing red. There are 26 people in our class altogether. We can say that 14 out of 26 people in our class are wearing red. We can also write this as a fraction, 14/26.

Write on the board or chart paper the fraction that describes the group of people who fit your rule and label it accordingly:

$\dfrac{14}{26}$ wearing red

Ask students what fraction would represent the number of people who do not fit the rule. Have a student write that fraction and label it:

$\dfrac{12}{26}$ not wearing red

Play Guess My Rule several more times, each time recording the results in terms of fractions. Try to vary the fractions that result from playing the game, so some of your rules result in fractions both below and above 1/2.

Fractions to Describe Our Class

After you have come up with a list of fractions, choose a pair of fractions from one game of Guess My Rule and examine their relationship. For example:

10/29 of the students in our class are wearing sneakers; 19/29 are not.

Ask a pair of students to make a Class Strip partitioned to represent each student in the class. Point out that it needs to have the same number of sections as people in the whole class. You may choose to have them tape a paper strip to the wall or draw one on the chalkboard. Have the student pair mark and label the two fractions of the class wearing and not wearing sneakers. They should use different marks, such as X's from the left and ✓'s from the right, for each fraction on the strip. The following diagram illustrates our example:

What do you notice about these two fractions?
Which one is a bigger part of the class? How do you know?
Is either of them close to half the class? How do you know?

(Students will probably notice that the two fractions add up to 1, since together they represent the entire class.)

Refer to the **Dialogue Box**, What Do You Notice About These Fractions? (p. 9) for examples of student observations.

Making Mystery Rules   Divide the class into groups of three or four. Each group writes down its own Mystery Rule describing some students in the class. Have each group secretly show you its rule so you can be sure the category can be readily understood by the others in the class.

❖ **Tip for the Linguistically Diverse Classroom**   Have limited English proficient students offer their ideas to the group by drawing rebuses or by pointing to examples.

Bring the class members together and have each group take a turn challenging the rest of the class to Guess My Rule. The groups might start by saying which members of their own group fit their rule and which members don't.

At the end of each game, students determine the fraction that describes the group of people who fit their rule and the group of people who do not fit their rule. A student from the group who made up the rule acts as the recorder, writing the fraction of people who fit the rule and the fraction who don't fit the rule. For example:

$\frac{9}{29}$ buttons on shirts

$\frac{17}{29}$ no buttons on shirts

Keep these data to use during the next several sessions.

Session 1 Follow-Up

Working with Two Rules at Once   Consider two rules you have already worked with—for example, "has brown hair" and "wears glasses." Explore the group of people who both have brown hair and wear glasses by having the people with brown hair stand at the front of the room and then sit down if they do not wear glasses. The people left are those who have both brown hair and glasses.

If the fraction of people who have brown hair is greater than ½ and the fraction of people who wears glasses is greater than ½, what could be the fraction of people who have both brown hair and glasses? Is it possible that this fraction is greater than ½? Less than ½?

 **Extension**

Guess My Rule is a classification guessing game in which players try to figure out the common characteristic, or attribute, of a set of objects. To play the game, the rule maker (you, a student, or a group of students) decides on a secret Mystery Rule. For example, classification rules for people might be everyone who is wearing orange or everyone who has brown hair.

The rule maker starts the game by giving some examples of people who fit the rule—for example, by having two students who are wearing orange stand up. The rule must be about something you can see. The guessers then try to find other individuals who might fit the rule: "Can David fit your rule?"

With each guess, the individual named joins the group that *does* fit the rule or the group that *does not* fit. Both groups must be clearly visible to the guessers so they can make use of all the evidence—both what does and does not fit—as they try to figure out what the rule might be.

Two guidelines are particularly important:

1. *"Wrong" guesses are clues that are just as important as "right" guesses.* "No, Rikki doesn't fit, but that's important evidence. Think about how Rikki is different from Tuong, Pinsuba, and B.J." Here is a wonderful opportunity to help students learn that errors are not just mistakes but can be important sources of information.

2. *When you think you know what the rule is, test your theory by giving another example, not by revealing the rule.* "Luisa, you look like you're sure you know what the rule is. We don't want to give it away yet, so let's test out your theory. Tell me someone who you think does fit the rule." Requiring students to add new evidence, rather than making a guess, serves two purposes: It allows students to test their theories without revealing their guesses to other students; and it provides more information and more time to think for students who do not yet have a theory of their own.

When students begin choosing rules themselves, they sometimes think of rules that are too vague (for example, "students with a piece of thread hanging from their shirts"). You can guide students in choosing rules that are "medium hard"—not so obvious that everyone will see them immediately but not so hard that no one will be able to figure them out. The students should be clear about who would fit their rule *and* who would not fit; this eliminates rules like "wearing different colors," which everyone will probably fit.

Keep the mystery and drama of the game high with remarks such as, "That was an important clue," "I think maybe Jesse has a good idea now," and "I bet I know what Marci's theory is."

It is surprising how hard it can be to guess what seems to be a very obvious rule (like "wearing green"). Teachers have found it is often difficult to predict in advance which rules will be more difficult than others. Sometimes it will be necessary to give additional clues when students are stuck. For example, one teacher chose "has buttons" as her Mystery Rule. After all the students had been placed in one of the two groups, still no one could guess the rule. So she moved down the line of students, drawing attention to each in turn: "I'm going to turn Kim around to the back, like this— see what you can see. You have to look really hard at Lesley Ann—look along her arms." Finally, students guessed the rule.

Since classification is a process used in many disciplines, you can easily adapt the game to other subject areas. Teachers have used Guess My Rule in social studies and science as well as for other aspects of mathematics. Animals, states, historical figures, geometric shapes, types of food, and countless other items can be classified in different ways.

As you and your students play this game, you will find yourselves becoming more observant and more flexible in your thinking about similarities and differences.

What Do You Notice About These Fractions?

This discussion takes place during the activity Fractions to Describe Our Class (p. 5).

Twelve out of 28 people have stripes somewhere on their clothes. Sixteen out of 28 people don't have any stripes on them.

The teacher writes on the board:

$\dfrac{12}{28}$ stripes

$\dfrac{16}{28}$ no stripes

What do you notice about these fractions?

Tyrone: More people don't have stripes. Both fractions have 28 on the bottom.

Irena: If you add 16 to 12, it makes 28.

Lina Li: Twenty-eight plus 28 is equal to 56.

Tyrone: They're all even numbers.

Nhat: Twenty-eight minus 16 is equal to 12.

The next two fractions are 27/28 (no purple sneakers) and 1/28 (purple sneakers).

Anyone want to say something about these fractions?

Emilio: Twenty-eight is the only even number.

Irena: Twenty-seven plus 1 is equal to 28.

Shiro: Twenty-eight is the only number on the board that appears twice.

Marci: Twenty-eight minus 1 is equal to 27.

What about the number 28? Shiro said he saw it twice here. Is it anywhere else on the board?

Lina Li: Yes. I see it twice up there, too, in 16/28 and 12/28.

Why do you think you see 28 so many times on the board?

Marci: Because it's 16 plus 12 and 27 plus 1.

Shiro: Because that's how many people there are in our class.

Emilio: And that number is all the students who we asked.

So if I told you that 18/28 [*writing on board*] of our class members have sneakers on, what fraction would not have sneakers on?

Lina Li: Ten people.

Nhat: Ten out of 28 people.

And how did you know?

Nhat: Because 28 minus 18 is equal to 10. And 28 is the number of students in the class.

At first, students notice both useful and irrelevant information. Gradually, as their observations accumulate, the teacher brings their attention to one of their most important observations—that 28 comes up in all the fractions. Through using and thinking about fractions in context, students are attaching meaning to the numerators and denominators of these fractions and are understanding how the two fractions represent the entire class.

Materials

- Strips of adding machine tape 18 inches long (3–4 strips per student)
- Interlocking cubes or counters (30 per student)
- Markers or crayons, scissors, tape
- Class Strips (2–3 sheets per student)
- Transparency of Class Strips
- Student Sheet 1 (2 per student, homework)
- Large fraction strip prepared in advance
- Family letter (1 per student)
- Overhead projector

Finding Familiar Fractions

What Happens

Students review the fractions ¼, ⅓, ½, ⅔, and ¾ by folding strips of paper into fractional parts. They use Class Strips to illustrate data about a group of people. By folding the Class Strips into halves, quarters, or thirds, they are able to describe such fractions as 14/26 in more familiar terms—for example, "a little more than ½." Their work focuses on:

- reviewing the size of familiar fractions relative to one another
- understanding the size of unfamiliar fractions by identifying which familiar fractions they are close to

Activity

Making a Fraction Strip

Pass out one 18-inch strip of adding machine tape to each student. Have students fold the strip in half, mark ½ on the fold, and draw a line *lightly* on the fold as on a ruler so they can see it more clearly. Make sure students do not label the area between the folds. In order to see the relationships among the fractions clearly, students need to label the folds, not the areas between them.

When you folded your strips in half, you ended up with two equal parts. How could you fold the same strip of paper so you end up with four equal parts?

Some students will suggest that you fold the strip in half and then in half again. Other students might suggest you fold the strip in half, open it up, and then fold each half in half by folding the ends into the center. Either of these methods will yield four equal parts.

Students now add fourths to their strips of paper by folding and labeling the folds. Check to make sure students are labeling the folds, not the areas between them. Students sometimes want to label each fold "¼" because they are thinking about labeling the areas. Be sure they label the folds ¼ and ¾. If a student makes a mistake, he or she can start a new strip by folding it in half first and then trying fourths again. Students may also observe that the ½ mark could also be labeled "2/4."

Note: Some students may call fourths "quarters" and others may not be familiar with this way of naming fourths. You might take a moment to talk about these two ways of naming fourths and make the connection with a quarter (the coin) as a fourth of a dollar.

Now give each student a new practice strip and have him or her fold it to make thirds. Ask students to share their methods. Folding strips into thirds may be difficult for some students. Students who have good methods can help others. When everyone has a way to fold a strip into thirds, have students fold into thirds the strips they already labeled with fourths and halves and to mark and label ⅓ and ⅔:

$$\boxed{\quad \frac{1}{4} \ \frac{1}{3} \qquad \frac{1}{2} \qquad \frac{2}{3} \ \frac{3}{4} \quad}$$

Ask students what they can see from their strips:

Which is bigger, ⅓ or ¼? How do you know? What fractions are bigger than ½? Do you know any other fractions that are bigger than ½? Where are 0 and 1 on this strip?

Post the large fraction strip you prepared in advance and keep it posted for the rest of the unit.

Estimating Familiar Fractions

Return to the data from the Guess My Rule games played in Session 1. Choose from one of the games a fraction that is close to a unit fraction (½, ⅓, or ¼) but not exact. Fractions such as 11/23 (close to ½) and 8/26 (close to ⅓) are good choices.

Note: If some students were absent when you collected the data and you want to include them, you can either add them to the data (if the categories are permanent ones, such as hair color) or quickly re-collect the data (if the categories are changeable ones, such as clothing color).

Yesterday, when we played Guess My Rule, we found out that 8/26 of the people in our class were wearing rings [*use relevant data from your* *class*]. If we were reporting our data, we probably wouldn't want to say 8/26, because it's not a fraction that's easy for most people to imagine.

If we could find a fraction like ½ or ⅓ that was close to this fraction, people would have a better idea of what part of the group we were talking about. When mathematicians or scientists do surveys and collect data, they often use "familiar fractions"—like the ones on your strip, ½ or ⅓ or ¾—that are close to the fractions they found. Using these familiar fractions also makes it easier to compare different groups.

If the fraction the mathematicians found in their survey was 8/26, what more familiar fraction do you think they might use to give people an idea of what they found? What familiar fraction is close to 8/26? You might want to use the strips we just folded to help you think about it. You can also use cubes if you want.

Give students a few minutes to work on this question. It is possible that only a few students will have any idea how to answer it. Some students may use their knowledge of multiplication and division or even addition and subtraction (8 + 8 + 8 = 24, which is close to 26) to make an estimate. Others may use the folded strips to reason about equal parts of the class. The **Dialogue Box**, Finding the Familiar Fraction for 15/21 (p. 16), contains examples of how students may approach these questions for the first time. If only a few students have any idea how to think about the problem, ask a few more specific questions:

Is 8/26 more or less than ½? How do you know? Is it more or less than ¼? How do you know?

Activity

Using Class Strips to Find Familiar Fractions

Here's a way to find familiar fractions that is similar to the paper folding you just did.

Pass out a sheet of Class Strips (p. 67) to each student. Students cut out an individual strip.

Note: There are 26 sections on each strip, so if there are fewer than 26 students in your class, students should cut off the extra sections. If there are more than 26 students in your class, students will have to tape two Class Strips together in order to have enough sections for every student. If students were absent on the day you collected the data, you will need to use the total of students that were in class that day. Or you might want to quickly re-collect the data so you can use the total number of students in class today.

Students then put an X or some other simple mark on a section of the strip (starting with the leftmost section and filling in consecutive sections) for every student who fits the rule you are discussing: "8/26 of the people in our class were wearing rings yesterday."

Which sections of your strip show the people who weren't wearing rings? How could this strip help you in figuring out what familiar fraction 8/26 is close to? Can you fold your Class Strip to see whether [your data] is more or less than 1/2? 1/4? 1/3?

See the **Dialogue Box**, Finding the Familiar Fraction for 15/21 (p. 16), for some examples of class conversations and the **Teacher Note**, Folding Familiar Fractions (p. 15), for some ways to support students' work with Class Strips.

Students help one another fold their Class Strips. If necessary, model the folding process for students. Remind them that the idea is to find a fold representing a familiar fraction so most of the marked sections are on one side of the fold and most of the unmarked sections are on the other side. You can demonstrate this on the overhead projector using transparencies of Class Strips that you mark and fold according to the fraction being analyzed.

Write on the board a statement that uses a familiar fraction and an approximating phrase (for example, "a little more than," "almost," "a little less than") to describe the data. Use a full sentence so students remember what the numbers refer to. You may have more than one such statement for one piece of data: "A little less than 1/3 of the people in our class were wearing rings yesterday." "A little more than 1/4 of the people in our class were wearing rings yesterday."

Choose from the Guess My Rule data another fraction that is close to a different familiar fraction. If you feel your class is ready, use a fraction that will be close to a nonunit fraction. Students often find the nonunit fractions, such as 2/3 or 3/4, more difficult than the unit fractions (see the **Teacher Note**, Folding Familiar Fractions, p. 15). Students use a new Class Strip to determine what familiar fraction is close to the actual data fraction. Some students may want to develop their own approach using cubes or other counters.

They share their conclusions. Write on the board a statement about the data, such as this: "A little more than 3/4 of the people in our class have brown hair."

If there is time, students work in pairs to find familiar fractions for two more fractions from the class Guess My Rule data and record them on Student Sheet 1, Finding Familiar Fractions (p. 57). Before the end of the session, each student finds in the class data two fractions for which he or she has not yet found familiar fractions and copies it onto a second copy of Student Sheet 1 for homework.

Session 2 Follow-Up

 **Homework**

Finding Familar Fractions   For homework, students find the familiar fractions for the data they recorded on Student Sheet 1, Finding Familiar Fractions. They record their results and explain their strategies. Each student should take home a sheet of Class Strips to use for this. Also send home the family letter or *Investigations* at Home booklet.

Folding Familiar Fractions

Some students may have trouble coordinating an unfamiliar fraction, a physical representation, and several familiar fractions with different denominators. In order to fold their Class Strips, students first have to have an idea what familiar fraction might be close to their data fraction—or they must fold all three familiar fractions (halves, fourths, and thirds) and keep track of all of them to see which is closest. Beginning with unit fractions is one way to simplify the task. Because students can usually do the arithmetic associated with unit fractions (fractions where the numerator is one), they have a way to figure out which kind of fold to try first. For example, suppose that $7/22$ of the class members were wearing watches. Some students will know that 7 is $1/3$ of 21 and is therefore close to $1/3$ of 22. Some students will fold over a chunk of 7 sections on the Class Strip, then fold another 7, and then fold another 7, and then count the number of sections (three, in this case) to get the familiar fraction. Others will fold the strip in thirds and see that they have about 7 sections of the strip in each third.

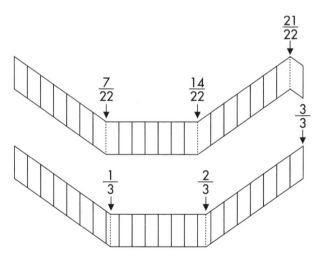

Using cubes is easier for some students. Students can divide 22 into thirds by dealing out cubes into 3 piles until they run out, then noting that there are 7 in each group and only 1 left. Or they might make groups of 7 until they run out, then count that there are 3 such groups with only 1 left over. Be sure you have cubes available for

students who find them easier to work with. You can model this approach to help students get started:

Do you think $7/22$ might be close to $1/2$? Try it and see—divide your 22 cubes in half. . . . Is $1/2$ too big? Try thirds—divide your cubes into thirds. Is $1/3$ of 22 close to 7 out of 22?

Having the image of the fraction strip on the board or on a large piece of paper is also helpful to students, as it reminds them of how to fold for familiar fractions, what order the fractions are in, and approximately how far apart they are.

If students fold their Class Strips several times to try out halves, thirds, and fourths, they may become confused about which fold corresponds to which fraction. In this case, they can mark the fold on their Class Strip with $1/2$, $1/3$, $2/3$, and so on so they can keep track.

Because finding a familiar fraction to express a more complex one is an estimation process, there is no single correct answer. The fraction $15/21$, for example, is between $2/3$ and $3/4$ and is approximately the same distance from each. Some students may estimate 21 is close to 20, so $15/21$ is close to $15/20$ or $3/4$. Others may estimate that 15 is close to 14, so $15/21$ is close to $14/21$ or $2/3$. Either answer is a reasonable approximation. Encourage this kind of reasoning, but realize that in many cases it will be beyond students' fractional knowledge to judge which estimate is closest to the actual fraction.

One mathematical caution: Because finding familiar fractions is a process of estimation, it is possible that the familiar fraction of the class who fits the rule and the familiar fraction of the class who does not will not add up to 1. For example, the familiar fraction for $17/28$ could be $1/2$ and that for $11/28$ could be $1/3$. But $1/3$ and $1/2$ do not add up to 1. This happens because $11/28$ is really a little more than $1/3$ and $17/28$ is really a little more than $1/2$.

Finding the Familiar Fraction for 15/21

This discussion takes place during the activity Using Class Strips to Find Familiar Fractions (p. 12).

How could this strip help you figure out what 15/21 is close to? What would you say it's closest to: 1/2, 1/4, 2/4, 3/4, 1/3, or 2/3?

Alex: You could fold it in half. It would tell you that it's at least 1/2.

Nick: You could count off different fractions like 1/4 is 5. Fold the fraction that 15 is closest to, like if you thought it was fourths you could count off 1, 2, 3, 4, 5 sections and fold it, then 1, 2, 3, 4, 5 more and fold it—that's 1/2, then 5 more—that's 15. So 15 is closest to 3/4.

Luisa: I'd divide 21 into 3 parts, so that's 7. Then I'd count off 7 and fold the strip, count off 7 more and fold. Then it's a little more than 2/3.

Kim: If you fold it in half, then there are 10 on each side of the fold. I'm not counting the one in the middle. So it's more than 1/2.

Qi Sun: I think it's 2/3 because 1/3 of 21 is 7 and 2/3 of 21 is 14, so 15 is just 1 off.

Did you use a strip to help you think about the problem?

Qi Sun: [*picking up the strip and pointing to thirds*] If each of these sections represents 7, then 2 sections is 14, so I'd say that 15/21 is about 2/3.

Sarah: I did it the way Qi Sun did it.

Tell us how.

Sarah: Well, some people might say it's in quarters because 5, 10, 15, 20, and 20 is close to 21, but I think of it in thirds, and it's a little more than 2/3.

Luisa: First I thought it was going to be this one [*she points to quarters*], but I think it's this one in thirds because I made 21 into thirds and 2/3 is 14 out of 21.

Pinsuba: I thought that it's going to be thirds because 21 is an odd number and there are 3 parts in thirds and 4 parts in fourths, so thirds has an odd number, too.

Nick: I thought of it in fourths because each part is worth 5 and 15 is 3 parts, three quarters. If it was 20 people, then 15 out of 20 would be 3/4.

Qi Sun: But 1/4 of 21 is different than 1/4 of 20, so it's different.

Comparing Data with Familiar Fractions

What Happens

Students use familiar fractions to represent data about themselves and their families and to compare themselves with the country as a whole. Their work focuses on:

- collecting and organizing data
- representing and comparing data using fractions

 Ten-Minute Math: What Is Likely?   During the remainder of this unit, try to do this activity three to four times with your students during any spare ten minutes you have outside of math class. You will need a clear container filled with two colors of blocks, beans, tiles, or some other material of similar size and shape. The first one or two times you do the activity, put in the container much more of one color than the other—for example, 18 red cubes and 2 yellow ones.

Show the container to students.

Students predict which color they will get most often if they draw ten objects out of the container.

Ten students each draw one object from the container, replacing after each draw.

Record the color of each object drawn.

Discuss what happened.

Try it again.

For full directions and variations, see p. 51.

For full directions and variations, see p. 51.

Materials

- Interlocking cubes or counters
- Markers or crayons, scissors, tape
- Class Strips (3–4 sheets per student)
- Student Sheet 2, pages 1 and 2 (1 per student)
- Transparencies of Student Sheet 2, pages 1 and 2 (optional)
- Clear container filled with two colors of objects of similar size and shape (Ten-Minute Math)
- Overhead projector

Begin by quickly collecting data by having students first raise their hands if they have a dog and then raise their hands if they have a cat. Record the data on the board or overhead in fractions, as in the example below:

Have a dog? $\frac{13}{25}$

Have a cat? $\frac{19}{25}$

Activity

Teacher Checkpoint

More Data for Class Strips

In pairs, students find familiar fractions for these two fractions and write how they found them. Then they write one sentence using familiar fractions comparing the fraction of people who have dogs with the fraction of those who have cats. Allow students to use Class Strips or cubes to help them find familiar fractions.

❖ **Tip for the Linguistically Diverse Classroom** Have limited English proficient students create sentences using rebus drawings and mathematical symbols. For example:

As you observe students working on this task, notice if they understand how the numerator and denominator of the fraction stand for the students who have dogs (or cats) and the students in the class, respectively. Do they make a strip of appropriate length? Do they understand how to mark their strips? Do they know how to fold strips into halves, thirds, and fourths? Do they understand how to compare the real data, marked on their strips, with the familiar fractions they get by folding? You can make similar observations about the students who are using cubes: Do they model the whole group and the parts of the group appropriately? Can they compare the fraction they're working with to half the class, one-third of the class, one-fourth of the class? If a significant number of students are having trouble, you may want to do a few more examples with the whole class.

See the **Teacher Note**, Problems with Wholes (p. 21), for examples of students' potential problems with finding familiar fractions.

Activity

Comparing Our Class with National Data

I have some data about people in the whole United States, and I'm wondering how this class compares with that data. Are we like the people who responded to the survey or are we different?

Give each student a copy of Student Sheet 2, pages 1 and 2, Comparing Class Data and National Data.

Conduct a survey of the class on the following questions:

Do you have four living grandparents?
Do you wear eyeglasses?
Do you often eat vegetables?
Do you think cigarettes are bad?
Are you the oldest child in your family?
Do you live in the same state where you were born?

Record the class data in fractions on the chalkboard or on transparencies of Student Sheet 2. Students copy the class data onto their own Student Sheets to use in class and for homework.

Question	Your Class Data	National Data	Compare the Data
Have four living grandparents?		about ¼ of Americans (age 8 to 17)	
Wear eyeglasses?		about ½ of Americans	
Eat vegetables often?		almost every American	
Think cigarettes are bad?		about ⅔ of American youth	
Are the oldest child in their family?		a little more than ⅓ of Americans	
Live in the same state where they were born?		a little less than ⅔ of Americans	

(Source of data: Weiss, Daniel Evan. *100% American*. New York: Poseidon Press, 1988.)

Making a Comparison Choose one question to start with. Students work in pairs and may use Class Strips, cubes, or some other model to compare themselves with the national data. Although many students will use familiar fractions, some will compare the data by starting with the national fraction and figuring out how many of their class it would take to equal that fraction. For example, in a class of 29, they may figure out that about 7 students would be ¼ of the class. So if more than 7 students had 4 living grandparents, the class would have a larger fraction of people than the national data. Be sure to ask such students how they knew how many students are in ¼ of the class.

Students discuss their results.

How did our class data compare with the national data? How do you know?

What familiar fraction was our class data close to? Was our data a little more or a little less than this fraction? How do you know?

Making Other Comparisons For the remainder of this session students work in pairs to compare the data on several more questions. Class Strips and cubes should be available to students as tools for figuring out familiar fractions.

Students will be interested in discussing possible reasons for the differences between their class data and the national data. One of the most important reasons for differences is that the national data is not about just fourth graders, but about a sample of the entire population. So, for example, half of the population wear eyeglasses, but a lower fraction of young people wear glasses.

Some students may have a hard time understanding that it makes sense to use fractions to compare different-size groups. If this issue comes up, have some discussion about it, but don't expect all students to understand this idea (see the **Teacher Note**, Comparing Groups of Different Sizes, p. 22).

Session 3 Follow-Up

 **Homework**

Comparing Class Data and National Data   Students finish Student Sheet 2, pages 1 and 2, Comparing Class Data and National Data. Give students a sheet of Class Strips to use as a tool if they want it.

Problems with Wholes

As the problems in this unit become more complex, students sometimes lose track of what is the whole and what are the parts they are considering. Recognizing the whole is a critical part of understanding fractions.

For example, one student became confused, forgetting that her Class Strip already represented the whole class. She began to think of the whole strip as a part of something larger:

Kim: I think it's ½. If you put two Class Strips together, then this one (she picks up the one she has marked off the data on) is ½. If you have three strips, then this one is ⅓, so it's ½ because it's not ⅓. If you had four strips, it would be ¼.

Here is the conversation of another group of students as they move in and out of clarity about what the whole is and what fractional part they are considering. They are trying to find the familiar fraction for 5 out of 20, the number of students in their class who are 10 years old (15 of the students in their class are 9 years old):

Karen: I think it's ⅓.

David: Yeah. One-third of our class is 10 years old.

How did you figure that out?

Karen: I did 5 times what is equal to 15. No, that's not what I did. I think I counted how many

5's are in the 15 and divided it. So I got three 5's, and one part is ⅓.

David: It's actually ¼ because you have to include all the 10-year-olds. There are 20.

Jesse: What fraction of the class is 9 years old?

David: Three-thirds.

Karen: No, ⅔. Wait, it is ¼, isn't it?

David: Yes, because 3 groups of 5 and then 1 more 5.

Karen: So ¾ of the group is 9 years old and ¼ is 10 years old.

Another confusion for students, when working with fractions of a group, is between the part of the group (for example, ¼) and the number of things in that part. Here, for example, a student thinks that ¼ is 4 sections of the strip, rather than 1 of 4 equal sections:

Tuong is trying to figure out a familiar fraction for ⁴⁄₂₂. He writes the numbers 1 through 22 in the 22 sections of his strip. Then he counts off 4 sections and folds the strip over and over, so there is a fold after every 4 sections. He ends up with a strip that is folded into fifths with two left over. He explains: "I was trying to get fourths and since there were 4 sections I folded on every fourth one." He made fifths but called them fourths.

Students may ask how many people are in the national data set because they feel it is critical to know the number of people in each group in order to make a valid comparison. These students are probably thinking about comparing the *number* of people rather than comparing the *fraction* of people. Here are two ways to address the issue.

The population of the United States is so large that just about any fraction of it will be much larger than the number of students in your class. The population of the United States is about 250,000,000 (256,566,000 is the estimate from the U.S. Census Bureau for the population as of January 1993). So even one-fourth would be about 60 million. Comparing by number in this case just wouldn't make sense. We're not trying to see whether our class has more or fewer people who own dogs than in the whole country—we already know it's fewer! Rather, we're trying to see whether our class as a whole is similar to the country as a whole when we compare how big a part of our class owns dogs to how big a part of all Americans own dogs.

In Investigation 2, however, students compare their class with another class in their school, and the class sizes are likely to be close. In this case, some students may have an even harder time understanding why we use fractions rather than

numbers to compare groups. The following example may be helpful. Suppose I collect data from two clubs about dogs. In one club, 50 out of 100 people own dogs. In another club, 5 out of 10 people own dogs.

Question	Club 1	Club 2
Own a dog?	$\frac{50}{100}$	$\frac{5}{10}$

How can we use these data to compare the two clubs?

Which club has a greater fraction of dog owners?

Don't be surprised if some of your students don't believe you can compare groups if there are unequal numbers in the groups. The idea that you can compare groups by using fractions (or percentages or averages), rather than directly comparing amounts, is a complex idea. It requires an understanding that a group can be considered as one whole and can be compared to another group also considered to be one whole. This idea requires that our view of the group change: We no longer consider size as an important attribute; we view each group as a unit. Much older students still have trouble with this idea, so don't expect all your students to understand it completely.

Using Fractions to Compare Data

What Happens

Students discuss and compare their class data and the national data using their work from the previous session. Students do word problems that require them to relate fractions to real situations—for example, "How many people are in three-quarters of the class?" Their work focuses on:

- comparing data from two different-size groups
- using data and fractions in word problems

Materials

- Cubes and Class Strips remain available
- Student Sheet 3 (1 per student, homework)
- Overhead projector (optional)

Activity

Discussing Findings About the Data

If you did not discuss some of the data comparisons students did yesterday, discuss them now, and include the comparisons done for homework. Focus both on the techniques students used to make the comparisons and on the reasons your class data may be different from the national data.

Activity

Using Fractions in Word Problems

As an introduction to using fractions in word problems, ask students:

Do you agree with the following statement? Three-fourths of the people in our class have brown hair.

Give students a few minutes to work on this problem in pairs. When they are ready, ask them how they thought about it. They can use cubes, Class Strips, or pictures to solve the problem. They should not only figure out the answer but also be able to explain how they solved the problem.

What are the steps you went through to determine if the fraction statement is true or false?

Tell students that today they'll work on some problems similar to this one. Put the following word problem on the board or overhead.

> There are 24 kids in a class. One out of 3 kids plays on a soccer team. How many kids in this class play soccer on a team?

Students work on this individually and then with a partner. While they are working on the problem, observe students to find several different strategies they are using.

❖ **Tip for the Linguistically Diverse Classroom** Ensure that each word is comprehensible to limited English proficient students by using rebuses and real examples.

Showing Strategies for Solving the Problem Ask for volunteers to show at the overhead how they solved this problem. Encourage students with different strategies to share their ideas so students see that there is more than one way to think about the same problem.

Here are some strategies students used to solve the soccer team problem: One student made a row of 24 cubes, divided it into groups of 3, and counted the groups. Another student lined up 24 cubes, marked 1 of every 3 (every third one), and counted them to get 8. A third student working with cubes moved 1 of every 3 cubes away to make a pile of 8. Other students used pencil and paper, making 24 marks on the page and circling every third one. In a slightly different approach, some students made groups of 3, which added up to 24, then counted the groups. A few students used Class Strips of length 24, which they folded into thirds, then counted the number of sections in one-third.

When the class is ready, try a new problem:

> Three out of 4 people in a fifth grade class like spaghetti. There are 24 people in the class. How many people in the class like spaghetti?

Writing about the Problem This time students work in pairs to discuss and solve this problem. Then each student writes about how he or she solved the problem.

❖ **Tip for the Linguistically Diverse Classroom** Have limited English proficient students show you how they solved the problem.

If you have more time remaining in this session, students can work on similar problems on Student Sheet 3, Word Problems.

Word Problems Students finish the problems on Student Sheet 3. For some students, you may want to assign only one or two problems. Give students a sheet of Class Strips to use as a tool if they want.

Homework

INVESTIGATION 2

Looking at Data in Categories

What Happens

Session 1: Games We Play Students think of games they like to play and find ways to categorize them. They make graphs of the data using different categorizations and describe what they can see from their graphs. Students are encouraged to use fraction statements as part of the description of their data.

Session 2: More Games, and What Have We Eaten? Working as a whole class, students make graphs of their favorite games, using two different sets of categories. Then in small groups of two or three, students devise categories and graph foods they've eaten in the last 24 hours.

Session 3: What Do You Want to Be When You Grow Up? Students collect data about what they would like to be when they grow up. Working in small groups, students sort ten of these responses into categories that they make up and predict whether data from the first grade will be different from their own. They make plans to collect first grade data before the next session.

Session 4: Organizing Some First and Fourth Grade Data Students make data cards with the career data from ten of the first graders. They combine these cards with the ten data cards they already categorized from the fourth grade. As they add the first grade data, students make the necessary adjustments in their categories to accommodate the new data. Then groups of pairs meet to share how they made categories and to discuss what issues came up in doing this.

Sessions 5, 6, and 7: Making Comparisons with All the Data Students work with the complete set of career data from the first and fourth grades. Starting with the categories they made for the first 20 pieces of information, they categorize the remaining data. Some students will find they need to make additional categories or may want to

combine categories. Students graph their data and write about their observations. They also compare the fraction of fourth graders and the fraction of first graders in some of the categories. Finally, students share conclusions about the data based on their graphs.

Mathematical Emphasis

- Collecting and recording categorical data
- Organizing data into categories
- Making judgments about sets of categories
- Redefining categories to accommodate additional data
- Representing categorical data, including use of bar graphs
- Describing categorical data
- Using fractions to compare categorical data from two groups

What to Plan Ahead of Time

Materials

- Interlocking cubes, counters, and Class Strips remain available (Sessions 1–7)
- Index cards and tape, or large stick-on notes: several per student plus 10 extra (Sessions 1–7)
- Chart paper (Session 1)
- Crayons or markers, glue, scissors (Sessions 1–7)
- Large paper (11-inch by 17-inch) for making data displays: 2 per pair (Sessions 2, 5–7)
- Paper clips: about 10 per pair (Sessions 3–4)
- Envelopes: 1 per pair (for storing Data Cards, Session 3)
- Blank paper for labels: about 5 per pair (Session 4)
- Overhead projector (optional)

Other Preparation

- Duplicate student sheets and teaching resources (located at the end of this unit) in the following quantities. If you have Student Activity Booklets, no copying is needed.

  For Session 1

  Data Cards (p. 65): 6–8 sheets per student for use in Sessions 1–7

  Student Sheet 4, Games People Played (p. 61): 1 per student (homework)

  For Session 2

  Student Sheet 5, Transportation Data (p. 62): 1 per student (homework)

  For Sessions 5–7

  Making Comparisons with All the Career Data (p. 66): 1 per student

  Student Sheet 6, Comparing First and Fourth Graders (p. 63): 1 per student

  Student Sheet 7, What Do You Want to Be When You Grow Up? (p. 64): 1 per student (homework)

- Arrange with one of the first grade teachers for your students to collect career data from first graders between Sessions 3 and 4. You will want data from one first grade class (or about 20 to 30 students). Don't collect data from more than one class, unless the classes are extremely small, or your students will have a hard time managing all the data. Some first grade teachers will want to use the data for activities of their own, so you may decide to make two copies of the data. First grade teachers may also be interested in having the fourth graders share their findings with the first grade class after they have analyzed the first grade data in comparison to the fourth grade.

- Make two clear lists, one of all the first grade career data and one of all the fourth grade career data not yet used by your students in making their categories in Sessions 3 and 4. Either post a few copies of these lists where all students can see them in order to copy the data onto Data Cards or duplicate the lists so each student pair will have a copy of the first grade and fourth grade data. (Before Session 5)

Games We Play

Materials

- Data Cards (1 per student)
- Scissors
- Index cards and tape, or stick-on notes: about 10 (optional)
- Chart paper
- Glue or glue sticks
- Large paper (1 per group of 2–3 students)
- Student Sheet 4 (1 per student, homework)

What Happens

Students think of games they like to play and find ways to categorize them. They make graphs of the data using different categorizations and describe what they can see from their graphs. Students are encouraged to use fraction statements as part of the description of their data. Their work focuses on:

- collecting and recording categorical data
- organizing data into categories
- graphing categorical data

What Games Do You Like?

First, brainstorm for a few minutes with students about games they play. Encourage students to include all kinds of games: board games, games they play at school, computer games, math games, outdoor games, team games, card games, and so forth. If students at first think only of certain kinds of games (for example, only sports or only board games), ask questions to expand their thinking:

Are there any games you like to play that are outdoor games? Can you think of any games you play at school? Are there games you like to play with your family?

The purpose of this brief brainstorming is just to help students think about the broad range of games they play. It need last only a few minutes.

Data Collection Now each student thinks of up to five games he or she likes to play and writes the name of each game in one rectangle on a sheet of Data Cards. Let students know they can choose any kind of game and can fill in less than five rectangles if they can't think of that many games they like. Students can choose some of the same games, but encourage each student to make his or her own decisions (rather than, for example, a group of three students all agreeing on the same five games to write down individually).

Students should print the names of the games clearly, then cut out the rectangles they've used. Save the rest of the Data Cards for the next data activity in Session 2; they will use more of the rectangles then.

What Can You See from a Graph?   As students are writing and cutting, choose two who will not be working together in the same small group and who have a little bit of overlap in their selections of games. Sketch a graph of their selections on the board:

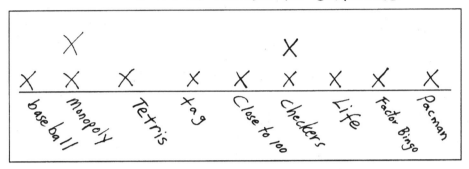

Note: If you can quickly write the game names on index cards with a loop of tape on the back of each, or on large stick-on notes, you will be able to move them around on the board quickly as you make different graphs in the upcoming discussion.

When most students are finished writing and cutting, ask everyone to look at the graph you've made:

What can you tell from this graph about the games Vanessa and Emilio like?

You will probably get a lot of responses about each individual game: "One person likes Tetris," "Both of them like checkers and Monopoly," "One of them likes baseball." On chart paper, start making a list of what students observe. After students have made some observations, ask them to categorize these games in a different way.

When mathematicians see a graph like this with lots of individual items, they often try to put the data into categories to see what else they can learn about the data. Let's see if we can learn more about our favorite games by grouping some of our data. Does anyone see a way to group these games? Are there some that go together?

Following students' suggestions, make a new graph, such as the one on p. 30, that groups the games in some way.

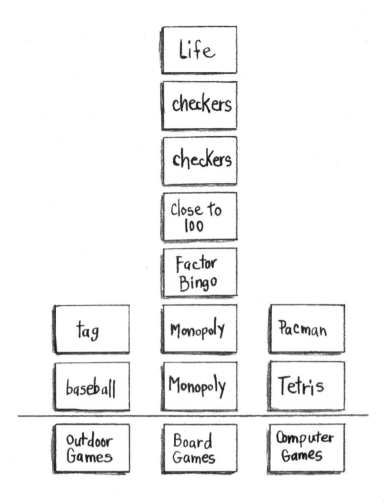

Now what can you tell from this graph?

On your list, record what students have to say. This time, students might say something like, "Vanessa and Emilio like board games more than any other kind of game."

Ask students to try to think of a different way they could divide these games into categories. If students have a hard time thinking of a different way, you might make a graph listing the games for a different categorization, such as games played indoors and games played outdoors or games played with a lot of people and games played with a few people. Ask them to guess what categories you have in mind.

Deciding on Categories and Graphing the Data In groups of two or three, students combine their own data to make a graph. We have found that this is plenty of data for students to work with during this first attempt to group their data. Each group should try at least two different ways of categorizing its data. Then they decide which one they want to use and make their graph by pasting their rectangles on large paper. Each group writes a few sentences on its paper describing at least three things the group members can see from the graph.

Games People Played   Give each student a copy of Student Sheet 4. Students interview at least three adult friends or family members about games they played when they were children. Students then write observations about ways in which the information collected from these adults compares with the class data.

Homework

Games from Different Cultures   Begin a collection of games from the cultures represented in your classroom. Encourage students to teach games they play in their neighborhood or family.

Extension

More Games, and What Have We Eaten?

Materials

- Index cards and tape, or large stick-on notes (1 per student)
- Scissors
- Glue or glue sticks
- Large paper (1 per group of 2–3 students)
- Data Cards (leftovers from Session 1 and extras)
- Student Sheet 5 (1 per student, homework)

What Happens

Working as a whole class, students make graphs of their favorite games, using two different sets of categories. Then in small groups of two or three, students devise categories for and graph foods they've eaten in the last 24 hours. Their work focuses on:

- categorizing data
- graphing categorical data

 Ten-Minute Math: What Is Likely? Continue to do this activity during any spare ten minutes you have outside of math class. Vary the ratio of colors of objects in the container or the kind of objects you are using. For full directions and other variations, see p. 51.

Activity

Categorizing Favorite Games

As class starts, ask each student to choose one of his or her favorite games and to write it on an index card with a roll of tape on the back or on a large stick-on note.

❖ **Tip for the Linguistically Diverse Classroom** Have all students include a visual clue on each card to ensure comprehension of the written words. For example:

> Checkers (drawing of a red and black game board)
> Basketball (drawing of a basketball and a hoop)

If limited English proficient students do not know how to say or write the name of their chosen game, have them just do the drawing. Then, have an English proficient student write the name of the game on the card.

In the meantime, begin to make a graph on the chalkboard by writing only the categories one of the student groups chose for its graph in the last session. As they finish, students attach their cards to the chalkboard in the space above the appropriate category to make a graph. An example is shown on the next page.

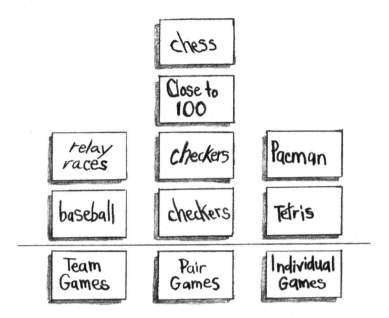

If some of the students' games aren't easy to categorize, put these to the side for now, then use them as examples for discussion about revising the categories. You might say for the above categorization:

Qi Sun is saying that Monopoly doesn't really fit into any of these categories because it's not a team game but you play it with more than two people. Qi Sun is also saying that tag doesn't really fit because you play it with a lot of people, but there aren't really teams. What should we do to make better categories so everything will have a place?

Once the categories are revised, ask students to review the data and list any observations on the board or a large piece of paper. Ask students to reflect on whether the categories help them find interesting information from their data. See the **Teacher Note**, What Makes a Good Categorization? (p. 36), for ways to support students in using categories to get useful information from their data.

After you have collected their descriptions of the data, ask students to begin to quantify some of their observations by using fractions. You might say:

Joey said that most students chose pair games as their favorite kind of game. What else can you say about that? About what fraction of the class chose pair games? Is it more than ½? Is it close to ¾? How do you know?

A few at a time, students retrieve their cards or stick-on notes from the board. Decide with the class on a different set of categories. You might use another set of categories that one of the groups made up in the last session. Write these on the board. Then have students come up a few at a time and place their cards in the appropriate category to make a new graph. Again, use any games that are difficult to categorize in the new scheme to discuss how to revise the categories so all the games fit somewhere. Then ask students to describe what this graph tells them about the data and to use fractions to quantify their observations when appropriate.

Teacher Checkpoint

Foods in the Last 24 Hours

Introduce the next activity:

You're going to be making graphs of the foods you've eaten in the last 24 hours. If we use the last 24 hours, when would we start? Which meals would be included? Who can remember what they had for breakfast this morning? For dinner last night? For lunch yesterday? How about any snacks?

After you are sure students understand the period of time you'll be considering, tell them they will each be selecting only seven of the foods they've eaten in the last 24 hours. They try to select foods they have eaten at different times during the day. For example, students might choose two foods from breakfast, two from lunch, two from dinner, plus a snack. They'll work together in pairs or threes to decide on categories for their foods and to make a graph, as they did with the games.

Working in groups of two or three, students choose seven foods eaten in the past 24 hours and write them on the rectangles on a sheet of Data Cards, then cut out the rectangles. The group members pool their foods and try grouping them in several ways. They choose the categories they want to use for their graph, make their graph, and again write at least three things they can tell about the data from their graph.

❖ **Tip for the Linguistically Diverse Classroom** Make sure groups represent a mixture of limited English proficient students and English proficient students. Have limited English proficient students draw the foods they cannot describe by writing. Then, have English proficient students add the written words. Encourage limited English proficient students to offer their ideas about what they can tell from the data on their graphs by creating sentences that include visual clues with or without words.

As you observe students' work, notice whether they are flexible in their thinking about the data. Can they think of several different ways to group the data? Are they beginning to understand that they can get different information depending on how they categorize the data? Do they make parallel categories that help them understand their data? Can they describe their data? Are they able to use fractions to quantify their description of the data? Your observations will help you determine what to emphasize and what kinds of questions to ask as students work on collecting data about careers in the next few sessions.

Session 2 Follow-Up

Homework

Transportation Data   Give each student a copy of Student Sheet 5, Transportation Data, to work on at home. Students may choose to use fraction strips, counters, or anything else that helps them determine the familiar fractions in the data. They explain their reasoning in writing. (Give students a sheet of Class Strips to use as a tool, if they want.).

❖ **Tip for the Linguistically Diverse Classroom**   To respond to each "How do you know?" section, have limited English proficient students attach the fraction strips they used to solve each problem. Tell them to make sure the strip shows how they used it (where they folded it, if they marked anything off, and so on). Students can also draw their explanations.

Extension

Games That Use Math   An interesting categorization for the games is "games that use math" and "games that don't use math." Make a list of the games students chose as one of their favorites at the beginning of Session 2; then duplicate the list for the class. Ask students to divide these games into the two categories (they can cut them out and do this in pairs at their desk or for homework). Then have a discussion about why they made the choices they did. This grouping can provoke a lively discussion about what mathematics is and where it comes up. Challenge students to be specific about why they think a game does or does not use mathematics. If there is one game that is particularly controversial, you can also have students write about their opinions.

What Makes a Good Categorization?

What makes a good set of categories for data? We use categories to help us see the data better, to reveal aspects of the data that may not have been apparent. Keep in mind that the reason for classifying data is to find what might be of interest in the data. Some classifications, while they would work, would not tell us anything about the data. For example, you could classify the games according to what letter each starts with, but this classification wouldn't give you good information about what games students like. Help your students keep in mind this question: "Do these categories help us learn anything more about the data?" Even reasonable categories might not reveal any new ideas about the data. When mathematicians look at data, they look at it in many different ways because they don't know in advance which way of viewing the data will be of most interest. Encourage your students to try several different categorization schemes until they find one that shows something interesting.

You can introduce the term *mode* into the conversations you have with your students about their data. The mode is the data value that occurs most frequently. For example, in the first games graph on p. 29, there are two modes, Monopoly and checkers. Each has two pieces of data:

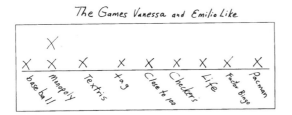

The Games Vanessa and Emilio Like

It is important that your students don't just say, "The modes are Monopoly and checkers," but that they consider whether the mode shows anything important about the data. Whether the mode is significant or not depends on how the data are categorized. In this graph, where every game has one or two pieces of data, the modes do not give us much information. But when

these data are categorized according to types of games (p. 30), we see that the mode is "board games." In this case, the mode is quite striking: there are 7 board games and only 2 games in each of the other categories. Classifying the data has given us a different look at what is going on.

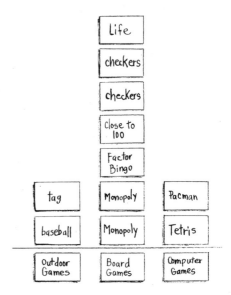

Note: Other statistical measures, such as *median*, cannot be used for categorical data because these data cannot be ordered by numerical value. A median is the middle of the data, but these data have no middle: They could appear in any order. Similarly, you cannot talk about the *range* of the data from lowest to highest values or where the data seem to cluster. These data do not behave like numerical data such as heights, ages, or how many people are in our families (see the unit *The Shape of the Data* for investigations and information relevant to ordered, numerical data).

A subtler idea that you can help your students with is creating categories that are related. For example, you could divide the people in your class into three categories: students who wear glasses, students who are nine years old, and students who are wearing red. This kind of classification gives you overlapping categories that

Continued on next page

are probably not related to one another. This can be a legitimate kind of analysis (you may be familiar with using Venn diagrams to look at intersecting characteristics) but is not the kind we are doing here. Usually, when analyzing categorical data, we try to choose a set of categories in order to answer a particular question about the data: Do students prefer indoor games or outdoor games? Do students like games that are played with a few people or many people? If students choose a set of categories that seem haphazard and unrelated, try to help them think about what question they might answer by using their categories.

Finally, it sometimes helps to break very large categories into some smaller ones in order to see information about the data. Breaking "indoor games" into several subcategories—perhaps "computer games" and "board games"—may reveal important aspects of the data that were not apparent before. If you see students with a very large and inclusive category, you might ask, "Do you see any way of dividing this category into two categories?" However, make it clear that while you want your students to try several different ways of classifying their data, they are in charge of deciding on their final categories.

What Do You Want to Be When You Grow Up?

Materials

- Index cards or blank pieces of paper (1 per student)
- Scissors
- Paper clips (5 per pair)
- Envelopes (1 per pair)
- Data Cards (1 sheet per pair)
- Overhead projector (optional)

What Happens

Students collect data about what they would like to be when they grow up. Working in small groups, students sort ten of these responses into categories that they make up and predict whether data from the first grade will be different from their own. They make plans to collect first grade data before the next session. Their work focuses on:

- collecting data
- organizing data into categories
- thinking about possible differences in categorical variables between the two groups

---

Activity

Collecting and Organizing Fourth Grade Data

In the next few days, we're going to be collecting data about what careers and jobs interest you and how your ideas about what you want to be when you grow up compare with what first graders think they want to be. First, we're going to collect our own data. Imagine yourself as an adult, and think about what you would like to do for a living. Maybe it is something you dream about or something you are already doing.

As students are thinking, hand out to each an index card or blank piece of paper. Students write down their answers to this question—"What do you want to be when you grow up?"—without discussing their answers with other students. There is no need for students to put their names on their responses. However, if anyone is absent, you will want to put aside cards for those students to fill out in the next day or so to add to the class data.

❖ **Tip for the Linguistically Diverse Classroom** Have limited English proficient students draw their answers.

Collect the students' responses, choose ten cards randomly from the fourth grade data, and record them on the board or overhead. Save the remaining fourth grade data for use in Session 5.

❖ **Tip for the Linguistically Diverse Classroom**   Next to each career written on the board or overhead, include a picture or drawing that relates to the profession. For example:

> Ballet Dancer (ballet shoes)
> Gardener (hoe)
> Police Officer (badge)

Organize the class into pairs and hand out one sheet of Data Cards to each pair. Students work in the same pairs analyzing career data during the next few sessions. The results from the Teacher Checkpoint in Session 2 may be helpful in pairing students who are confident working with categorical data with those who need additional help.

Each pair makes a set of ten data cards by printing the careers from the list on the board or overhead onto the rectangles on a sheet of Data Cards, then cutting them out. In addition, students write a small "4" or "fourth grade" in the corner of each rectangle, so that later they will be able to distinguish between the groups when they add first grade data.

❖ **Tip for the Linguistically Diverse Classroom**   Have all students include the same pictures from the master list on the individual data cards.

When the cards are complete, each pair thinks of ways to organize the data in categories. Students will probably need to read through the individual cards a few times in order to come up with a set of categories. Encourage students to talk with other pairs near them but to come up with their own approach to categorizing the data. See the **Dialogue Box**, Making Categories (p. 41), which is an example of two pairs of students discussing how to categorize their data.

When they have decided on a way to categorize the data they like, students write down on paper the names of the categories and the data underneath each category. They will use this list as a reference in the next session when they work with additional data.

After recording one way to categorize the data, students try to categorize them in another way, if possible, and write these categories on paper. This may be difficult with only ten data points. As students work with larger quantities in the next few days, they will have a few more chances to experiment with categorizing these data in different ways.

When pairs have agreed on one set of categories they think works best, they make a label for each of their categories (using scraps of paper or index cards) and paper clip each category label with the data cards that belong in that category. Students need a good place to store the cards (such as in an envelope) so they can continue working with them in the next two sessions.

Preparing to Collect First Grade Data

When each pair has recorded one way to organize the data, ask students to share some of the sets of categories they came up with.

Do you think the data from these ten fourth graders is typical of the whole class? What other categories might we need when we look at the whole class?

Then have students think about comparing their data to first grade data:

If we asked first graders what they want to be when they grow up, how do you think their answers might be different from ours?

In our experience, fourth graders have quite a few ideas about how first graders might think about careers differently from fourth graders. Students may mention specific careers at first. If this is the case, encourage them to think about more general trends as well. For example, if students think a higher proportion of first graders will want to be firefighters or police officers, would they predict more first graders will choose careers that have to do with public safety?

Before the next session, students need to collect data from first graders about what they want to be when they grow up.

How should we organize the data collection so that we get accurate information from the first graders?

Students will probably suggest a variety of ideas. Aspects of the data collection for students to consider before finalizing the plan are these:

1. The first graders must answer the question individually, preferably before they have had a chance to discuss their ideas.
2. Some first graders may have trouble reading and writing.
3. The data need to be collected before the next session, each recorded carefully on a data card, and returned to you or a central place you designate.

Some classes have decided to have each fourth grader interview a first grader individually. Since the numbers of fourth and first graders are usually not the same, some fourth graders may have to interview two first graders or a few fourth graders may team up to interview one first grader.

Note: For the purpose of comparing fractions later in the unit, it is mathematically more interesting if the total number of first graders interviewed is different from the total number of fourth graders, so don't worry about matching up your students one-to-one with first graders. Rather, it is better to interview one whole first grade class.

Making Categories

Two pairs of students are working at the same table as they categorize their ten pieces of data in the activity Collecting and Organizing Fourth Grade Data (p. 38).

Lesley Ann: Does "daredevil" go with "Performers" or "Sports People"?

Rebecca: I think a daredevil is more a performer than a sports person, because the daredevil is always thinking up a new act.

Lesley Ann: Or it could go separately with "mountain climber" under "Dangerous Jobs." The way I'm doing it is to look at people and see what is the same.

Tuong: Do we have to have the same groups as other pairs?

No, you can organize the data any way that makes sense to you. Each group may have a different way. There are a lot of ways to think about these data.

Rebecca: I'm not sure where to put "aerobics instructor." Maybe it should be in its own category.

Lesley Ann: It could go with "Teacher" or "Sports People."

Rebecca: I don't want to put it with "Teacher" because I don't think aerobics instructors really teach; people just follow what they do. We need a category of jobs that could be different things. Make up something in the middle that fits both.

B.J.: We were going to do "People Who Work Outside" and "People Who Work Inside." But scientists sometimes work inside; it depends what kind they are. Everybody goes inside sometimes.

Lesley Ann: I'm stuck about the engineer. Does that person want to be a train driver or a scientist?

Organizing Some First and Fourth Grade Data

Materials

- Paper clips (about 5 per pair)
- Scissors
- Data Cards (1 sheet per pair)
- Blank paper for labels (about 5 per pair)
- Overhead projector (optional)

What Happens

Students make data cards with the career data from ten of the first graders. They combine these cards with the ten data cards they already categorized from the fourth grade. As they add the first grade data, students make the necessary adjustments in their categories to accommodate the new data. Then groups of pairs meet to share how they made categories and to discuss what issues came up in doing this. Their work focuses on:

- constructing categories for categorical data
- comparing different sets of categories
- communicating their system for categorizing the data to other students
- looking at the same data categorized in a few different ways

Activity

Combining First and Fourth Grade Data

Before working with the first grade data, discuss briefly what it was like to collect data from the first graders.

Was any information hard to get? Were there any surprises? Did the first graders seem interested in the data we were collecting?

Choose ten first grade data cards randomly and make a list of the responses on the board or overhead. Save the remaining first grade data for use in Session 5.

Hand out one sheet of Data Cards to each pair of students, who then record the ten first grade responses, one response per rectangle. Before cutting them out, students label the first grade cards with "1" or "first grade" in the corner, to distinguish them from the fourth grade data. This is an important step, critical to the data analysis.

Categorizing Data Working in pairs, students refer to the list of categories they made to organize the ten fourth grade data cards and try to add the first grade data cards to these categories. Students will need to spread their first and fourth grade data cards out on tables, the floor, or several desks so they can see all their cards at once to organize them.

In order to include all the data, students may need to add or delete categories, expand the definitions of certain categories, or make the differences between some categories clearer. Some students may choose to start from scratch with a new system for organizing the data in categories. In most cases, this probably won't be necessary. Making clear labels for each category will make it easier to keep track of the different categories.

As you circulate, ask questions to support students in defining their categories clearly. If some students are making a lot of categories that have only one piece of data in each, encourage them to think of ways to combine some of their categories.

When pairs have agreed on a set of categories and decided where each of their Data Cards fits, they paper clip the data from each category to a category label.

Sharing Categories in Small Groups

Two or three pairs meet to tell how they made categories and to describe some of the decisions they made about specific pieces of data.

Some of you may find you had very different approaches to sorting the data. This is what makes working with categorical data interesting— there are many different ways to think about it. Each time we categorize the data in a different way we learn something new about data.

Discuss in your groups how you sorted the data. Tell group members what information can be learned from your set of categories.

Each pair explains their rationale so other students can understand what they did. As you listen to these conversations, encourage students to ask each other why they made the decisions they made. Students keep the categories they developed unless they find they can't explain why they put some of their data in the categories they did; then they may want to revise their categories. See the **Dialogue Box**, Sharing Categories (p. 44), for an example of a conversation between two pairs of students.

Class Discussion About Organizing the Data   After students have had time to discuss their categories in their groups, bring the class members together for a brief discussion. Students talk about some of the decisions they made as they decided on their categories. If students don't raise the following points on their own, ask the class:

What was hard about sorting the data?
What did you do with data that seemed to fit into more than one category?
Were there any responses you didn't know what to do with?
Do you think you will need to make more categories when we add the rest of the first and fourth grade data?

At this point, it is more important to acknowledge students' questions than to come up with definitive answers. When students add additional data, they will find some of the issues they are facing, such as categories without many data, are easily resolved. At the same time, new issues will spring from the data. For example, students may find they have data that fit in more than one category.

Sharing Categories

Two pairs of students are comparing the categories they used to group the data from ten fourth graders and ten first graders in the activity Sharing Categories in Small Groups (p. 43).

Joey: Our categories are "Sports," "People Who Teach Things," "Entertainment," "People Who Help Build Things," "People Who Design Things," "Business People," and "People Who Weren't Clear or Don't Know What They Want to Be." We decided to make sure at least three people fit in each category. Our biggest category, "Sports People," has nine people in it.

Lesley Ann: Where did you put the aerobics instructor?

Joey: Under "People Who Teach Things." We almost put it under "Sports," but David doesn't think aerobics is a real sport.

Rebecca: Why are "People Who Build Things" separate from "People Who Design Things"?

Nadim: Because the carpenter and builders actually put the building up. Architects and cake decorators don't do any building, they just design.

Lesley Ann: Our categories are different. We have "People Who Deal with Books," "Teachers," "Performers," "Scientists," "Jobs You Have to Be a Woman to Do," and "People Who Design Things."

Nadim: Where did you put the carpenter?

Rebecca: Under "People Who Design Things." My dad does a lot of carpentry, and he designs the project before he starts building.

Joey: What jobs do you have to be a woman to do?

Rebecca: Mother, businesswoman, and midwife.

Nadim: We had "Mother" under "Teacher," because mothers teach a lot of things.

Joey: Where did you put all the basketball players?

Rebecca: See right here; they are under "Performers." People pay to see them perform at games so we put them there.

Joey: I think they are sports people first.

Nadim: But maybe they wouldn't make so much money if they weren't performers, too.

Making Comparisons with All the Data

Materials

- Making Comparisons with All the Career Data (1 per student)
- Lists of the fourth and first grade data to post or hand out
- Large paper for making data displays (1 per pair)
- Crayons or markers, glue, scissors
- Index cards or stick-on notes, and tape (1 per pair)
- Cubes, counters, and Class Strips remain available
- Data Cards (2–3 sheets per pair)
- Student Sheet 6 (1 per student)
- Student Sheet 7 (1 per student, homework)

What Happens

Students work with the complete set of career data from the first and fourth grades. Starting with the categories they made for the first 20 pieces of information, they categorize the remaining data. Some students will find they need to make additional categories or may want to combine categories. Students graph their data and write about their observations. They also compare the fraction of fourth graders and the fraction of first graders in some of the categories. Finally students share conclusions about the data based on their graphs. Their work focuses on:

- adjusting categories to accommodate all the data
- graphing the data
- using fractions to compare first and fourth graders' career choices
- communicating about their findings
- comparing their work with alternative approaches to classifying the data

Activity

Combining and Organizing the First and Fourth Grade Data

Hand out a copy of Making Comparisons with All the Career Data (p. 66) to each student. This sheet describes the work of the next three sessions and is a resource for students.

Post or hand out the lists of the remaining career data for both first and fourth graders that you prepared in advance. Working in pairs, students print the data clearly on a sheet of Data Cards, then cut the cards out. Students continue to mark the grade level lightly on the front of each card, so they will be able to tell apart the first grade and fourth grade data. Students add their new cards to the set of career data cards they worked with in the last session.

Note: If you haven't done so already, collect data from those fourth graders who were absent the day these data were collected and add the new data to the class list.

When they are finished writing and cutting, each pair double-checks to make sure the number of data cards is equal to the number of data items, or students, of the two classes combined.

Now students finish sorting all of their cards according to the categories they developed in Sessions 3 and 4. Most students will find they need to make adjustments to their categories again to accommodate some of the new data. If they are not sure which category a piece of data goes in, they need to think more about their criteria. They may need to redefine their categories slightly rather than add a new one ("I think we should put 'veterinarian' with 'Doctor and Nurse' because it's more medical; the ones in 'People Who Work with Animals' are more like people who train them or sell them"). Because students may be working with two or three times the data they did in the previous session, they will need time to finalize their categories.

Final Projects: Graphing, Describing, and Comparing Career Data

Students complete three tasks including (1) graphing their data, (2) writing about their findings, and (3) comparing the fractions of first and fourth graders in two of their categories, using Student Sheet 6. Some students may choose to make the graph first, compare the data using fractions, and then write about their findings. Others may decide to start a list of observations as they are making their graph, write about their list of observations, and then compare the data using fractions. Each student pair decides on their own approach. The three activities are explained in more detail below.

Graphing Career Data Students who have organized all their data in categories are ready to make a graph showing the data in each category. Students paste their data cards on large paper in a permanent display. They label the name of each category and the number of people in each category. Part of their job is to design their graph so that somebody looking at it can easily see which are the first graders and which are the fourth graders. (Students might use color, patterns, location, or some other method to do this.)

Writing Observations About the Career Data On a separate piece of paper, each pair of students writes a few paragraphs about its conclusions. This writing can be about anything the students have noticed about the data, including qualitative and quantitative observations. They might also comment on why they chose their categories and which careers were hard to categorize.

Comparing First and Fourth Graders Students complete Student Sheet 6, Comparing First and Fourth Graders. Students choose two of their categories. They refer to their graphs to determine the fraction of first graders and the fraction of fourth graders in each category. They record these fractions on Student Sheet 6, determine in each category which fraction is bigger, and explain how they can tell.

Students may use cubes, Class Strips, or any other method that helps them compare the fractions. In some cases, it helps to determine the familiar fractions to make comparisons. For other comparisons, the relative size of the numerators and denominators may be enough information to make a valid comparison. When the fractions are very close in size, using cubes or strips may help in making the comparison.

❖ **Tip for the Linguistically Diverse Classroom** Have limited English proficient students draw their answers to the question at the bottom of Student Sheet 6.

Sharing Conclusions About the Career Data

Students post their graphs and written conclusions for the class to see. Then student pairs circulate around the room to observe one another's work. They write any questions or comments about particular graphs on stick-on notes they attach to that graph (or use index cards and tape). Comments focus primarily on the content of the graphs and the observations written about the graphs, rather than on topics such as neatness or spelling.

When everybody has had a chance to see the graphs, each pair of students shares with the class one or two conclusions about its findings. The pair also addresses any questions or comments it received on the stick-on notes.

After students have shared their findings, have a brief discussion about the findings more generally:

Were there some conclusions that more than one graph supported? What surprised you? Are there other questions you have now that would need more research?

Assessment
Evaluating the Final Projects

You can assess students' learning during this unit by examining their final projects, including their graphs, written conclusions, and comparisons of the data using fractions (Student Sheet 6), as well as the comments and questions they make about other students' displays. Here are some of the questions you can consider about students' work:

■ Did students choose reasonable categories? Did they define their categories clearly? Are they able to give reasons for why a piece of data goes in a particular category?

■ Does their graph depict the categories clearly? Can you easily tell from their display which responses are from first graders and which are from fourth graders?

■ Do the students' categories help them understand the data? Are they able to find and describe interesting aspects of their data? Do they do more than simply count the number of pieces of data in each category? Are they able to use fractions to describe and compare the data? Do they use their categories to help them compare the career choices of first and fourth graders? Are they beginning to interpret their data—thinking about why their data might be the way it is?

■ In their comparisons that use fractions, can students explain clearly why one fraction is larger than another? Do students use familiar fractions or concrete models to help them compare fractions?

■ Are students able to make observations and ask questions about other students' work? Do they notice how different information emerges from the data, depending on how they were categorized?

Choosing Student Work to Save

As the unit ends, you may want to use one of the following options for creating a record of students' work on this unit.

■ Students look back through their folders or notebooks and write about what they learned in this unit, what they remember most, and what was hard or easy for them. You might have students do this work during their writing time.

- Students select one to two pieces of their work as their best work, and you also choose one to two pieces of their work to be saved. You might include students' final projects, including their graphs, written conclusions, and comparisons of the data using fractions (Student Sheet 6). This work is saved in a portfolio for the year. Students can create a separate page with brief comments describing each piece of work.

- You may want to send a selection of work home for parents to see. Students write a cover letter, describing their work in this unit. This work should be returned if you are keeping a year-long portfolio of mathematics work for each student.

Sessions 5, 6, and 7 Follow-Up

What Do You Want to Be When You Grow Up?   Give each student a copy of Student Sheet 7 to work on at home. Students ask adults in their families or adult friends what they would like to be when they grow up. (We've found that adults are quite willing to answer this question, too!) As a follow-up; you can ask students to share their adult data, categorize, and compare these data to the first and fourth grade data.

What Is Likely?

Basic Activity

Students make judgments about drawing objects of two different colors from a clear container. They first decide whether it's likely that they'll get more of one color or the other. Then ten students each draw one object from the container, replacing after each draw. You record the color on the board before each student puts the object back. Students then discuss whether what they expected to happen did happen. They repeat the activity with another sample of ten objects.

What Is Likely? involves students in thinking about ratio and proportion, and in considering the likelihood of the occurrence of a particular event. Ideas about probability are notoriously difficult for children and adults. In the early and middle elementary grades, we simply want students to examine familiar events in order to judge how likely or unlikely they are. In this activity, students' work focuses on:

- visualizing the ratio of two colors in a collection
- making predictions and comparing predictions with outcomes
- exploring the relationship between a sample and the group of objects from which it comes

Materials

- A clear container, such as a fishbowl, a large glass, or a clear plastic jar
- Objects that are similar in size and shape, but come in two colors (blocks, beans, tiles, marbles)

Procedure

Step 1. Fill the container with two colors of blocks, beans, or tiles. When you first use this activity, put much more of one color into the container. For example, out of every 10 tiles you put in the container, you might use 9 red and 1 yellow. Thus, if you used 40 tiles, 36 would be red and 4 would be yellow. Mix these well inside the container. Continue to use these markedly different proportions for a while.

Step 2. Students predict which color they will get the most of if they draw ten objects from the container. Carry the container around the room so that all students can get a good look at its contents. Then ask students to make their predictions. "What is likely to happen if we pull out ten objects? Will we get more yellows or more reds? Will we get a lot more of one color than the other?"

Step 3. Ten students each draw one object from the container, replacing after each draw. Ask a student, with eyes closed, to draw out one object. Record its color on the board before the student puts the object back. Ask nine more students to pick an object, then replace it after you have recorded its color. Record colors using tallies.

RED |||| ///

YELLOW //

Step 4. Discuss what happened. "Is this about what you expected? Why or why not?" Even if you have a 9:1 ratio of the two colors, you won't always draw out a sample that is exactly 9 of one color and 1 of the other. Eight red and 2 yellow or 10 red and 0 yellow would also be likely samples. Ask students whether what they got is likely or unlikely, given what they can see in the container. "What would be *unlikely*, or surprising?" (Of course, surprises can happen, too—just not very often!)

Step 5. Try it again. Students will probably want to try drawing another ten objects to see what happens. "Do you still think it's likely that we'll get mostly reds again? Why? About how many do you think we'll get?" Draw objects, tally their colors, and discuss in the same way.

Variations

Different Color Mixes Try a 3:1 ratio—3 of one color for every 1 of the other color. Also try an equal number of the two colors.

Continued on next page

Different Objects   Try two colors of a different kind of object. Does a change like this affect the outcome?

The Whole Class Picks   See what happens when each student in the class draws (and puts back) one object. Before you start, ask, "If all 28 of us pick an object, about how many reds do you think we'll get? Is it more likely you'll pick a red or a yellow? A *little* more likely or a *lot* more likely?"

Students Fill the Container   Ask students to help you decide what proportions of each color to put in the container. Set a goal—for example:

■ How can we fill the container so that it's very likely we'll get mostly yellows when we draw 10 objects?

■ How can we fill the container so that it's unlikely we'll get more than one red?

■ How can we fill the container so that we'll get close to the same number of reds and yellows when we draw 10 objects?

After students decide how to fill the container, they draw objects, as in the basic activity, above, to see if their prediction works.

Three Colors   Put an equal number of two colors (red and yellow) in the container, and mix in many more or many fewer of a third color (blue). "If 10 people pick, about how many of each color do you think we will get? Do you think we'll get the same number of red and yellow, or do you think we will get more of one of them?"

The following activities will help ensure that this unit is comprehensible to students who are acquiring English as a second language. The suggested approach is based on *The Natural Approach: Language Acquisition in the Classroom* by Stephen D. Krashen and Tracy D. Terrell (Alemany Press, 1983). The intent is for second-language learners to acquire new vocabulary in an active, meaningful context.

Note that *acquiring* a word is different from *learning* a word. Depending on their level of proficiency, students may be able to comprehend a word upon hearing it during an investigation, without being able to say it. Other students may be able to use the word orally, but not read or write it. The goal is to help students naturally acquire targeted vocabulary at their present level of proficiency.

We suggest using these activities just before the related investigations. The activities can also be led by English-proficient students.

Investigation 1

true, false

1. Tell students that you are going to create *true* sentences about things they can see. Nod your head after each statement. For example:

   My dress is red. Tuong is wearing tennis shoes. Luisa has brown eyes.

2. Tell students that you are going to create *false* sentences about things they can see. Shake your head after each statement. For example:

   My dress is black. Tuong is wearing cowboy boots. Luisa has blue eyes.

3. Continue creating a mixture of true and false sentences. After each statement, challenge students to respond either *true* or *false*.

Investigation 2

gameboard, board games, outdoor games

Try to introduce these words in the context of playing games on gameboards or playing outdoor games. For example, if students play games using gameboards during indoor recess, or at some other time during the day, introduce the words *board games* and *gameboard* during this time. If students play outdoor games during gym or outdoor recess, introduce the words *outdoor games* during or right after these experiences.

1. Pull out and identify the gameboards of several popular games. For example:

   This is the gameboard used to play Monopoly. This is the gameboard used to play checkers.

2. As you point to all of them, explain that each is an example of a board game since each is played on a gameboard.

3. Next, show students a football, a baseball, a soccer ball, and a tennis racket. Point outside as you explain that these items are used for outdoor games.

4. Ask students a few questions about outdoor games and board games. For example:

   Would you play an outdoor game or a board game on a rainy day? Which games are more interesting to watch, board games or outdoor games? Why?

foods, snacks, breakfast, dinner, lunch

1. Ask students to write down or draw pictures of foods they typically eat for breakfast, lunch, dinner, and snacks.

2. Point to a clock that shows 7:00 as you write 7 AM on the board. Tell students that people eat breakfast in the morning.

3. Show and identify pictures, or use real examples, of foods that are typically eaten for breakfast. American meals may differ from what your students are accustomed to eating. Also, include examples of "typical" breakfast foods eaten by your students.

4. Continue showing and identifying pictures of foods typically eaten for lunch, dinner, and snacks. Be sure to change the clock to show the hour that corresponds with when the meal is usually eaten.

this morning, yesterday, last night

1. Point to today's date on the calendar. Pantomime actions as you tell students several things that already happened in the morning. For example:

Continued on next page

This morning I woke up at 6:30, ate breakfast at 7:00, and drove to work at 7:30. The first thing we did in class this morning was. . . .

2. Point to the preceding day on the calendar and tell students some of the events that happened yesterday. For example:

Yesterday, I woke up at 6:35, ate breakfast at 7:10, and drove to work at 7:40. Yesterday, the first thing we did in class was. . . .

3. Point to the same preceding day on the calendar. Then, draw a moon in a dark sky on the board. Pantomime actions as you tell the students what happened last night. For example:

Last night I took a shower at 8:00, read a book at 8:30, and went to bed at 10:00.

4. Ask students to pantomime something they did this morning, yesterday, or last night. Have students guess what each student is acting out. For example:

Luisa, act out something you did last night. Ahmad, act out something you did this morning.

careers, jobs

1. Show and identify several pictures of people doing a variety of job-related tasks such as fire-fighters, bakers, doctors, police officers, office workers, teachers, and performers.

2. Ask students what kinds of tasks the people in these jobs do. Have students pantomime the different tasks.

3. Have students organize the pictures according to jobs that require uniforms and jobs that don't; jobs that are mainly outdoors and jobs that are indoors; and jobs that are interesting to them and jobs that are not interesting to them.

Blackline Masters

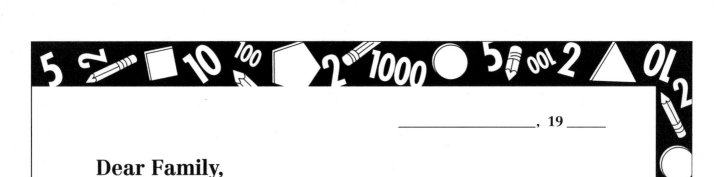

_____, 19 _____

Dear Family,

Our class recently started working on a math unit called *Three out of Four Like Spaghetti*. As the title suggests, this unit is about surveys in which the results are described and compared with other results by using fractions. Children will collect data about our own class—for example: What fraction of the students in our class eat vegetables often, wear glasses, or are the oldest in their family? Then we'll figure out what each of those fractions means. For example, is the fraction of oldest children in our class closer to one-third or closer to one-half? We'll then compare the fraction in our own class to the results of national surveys. For example, we'll find out how our class compares to the ⅓ of Americans who are oldest children. (Any predictions?)

Another thing we'll be doing is categorizing information that comes from surveys. We'll do our own class surveys of games children like to play, foods they like to eat, and jobs they want to do when they grow up. Rather than simply list results, we'll figure out how to categorize games or foods or jobs so we can better describe what we found. For example, children might categorize their favorite games by the number of players, by the amount of physical activity involved, or by where the games are played. This work of summarizing and categorizing is an important part of statistics because it allows us to make sense out of a lot of information.

In this unit, there are many connections to real life. You and your child probably run across many advertisements and newspaper articles with headlines like "three out of four doctors prefer X brand of pain reliever" or "½ of all Americans don't get enough sleep." Try to notice these ads and articles and talk with your child about them. "What does three out of four mean? What fraction is it? If there were 20 doctors, about how many of them would prefer this pain reliever? How do you think these data were collected?"

As we work with fractions and data, the children will be making pictures and diagrams, using strips of paper they divide into fractions, and working with other materials. When you work at home with your child, we suggest that he or she use pictures and objects to help solve problems. There are many different ways of making sense of data. At school, students will be encouraged to find ways that work for them. Encourage them to do this at home, too.

Sincerely,

Finding Familiar Fractions

Example:

Rule	Class Data	Familiar Fraction
Wearing red	$\frac{15}{28}$ *wear red*	*about $\frac{1}{2}$ wear red*
(⊞⊞ ⊞⊞ ⊞⊞) ⊞⊞ ⊞⊞ ∥∥ *28 is close to 30. 15 is $\frac{1}{2}$ of 30.*		

Rule	Class Data	Familiar Fraction
How did you determine the familiar fraction?		

Rule	Class Data	Familiar Fraction
How did you determine the familiar fraction?		

Comparing Class Data and National Data (page 1 of 2)

Question	Your Class Data	National Data	Compare the Data
Have four living grandparents?		about 1/4 of Americans (age 8 to 17)	
How did you decide which fraction is bigger?			

Question	Your Class Data	National Data	Compare the Data
Wear eyeglasses?		about 1/2 of Americans	
How did you decide which fraction is bigger?			

Question	Your Class Data	National Data	Compare the Data
Eat vegetables often?		almost every American	
How did you decide which fraction is bigger?			

Source of data: Weiss, Daniel Evan. *100% American.* New York: Poseidon Press, 1988.

Name _____   Date _____

Comparing Class Data and National Data (page 2 of 2)

Question	Your Class Data	National Data	Compare the Data
Think cigarettes are bad?		about 2/3 of American youth	
How did you decide which fraction is bigger?			

Question	Your Class Data	National Data	Compare the Data
Are the oldest child in their family?		a little more than 1/3 of Americans	
How did you decide which fraction is bigger?			

Question	Your Class Data	National Data	Compare the Data
Live in the same state where they were born?		a little less than 2/3 of Americans	
How did you decide which fraction is bigger?			

Word Problems

1. In a class of 28 students, 1 out of 4 students has braces. How many students in the class wear braces? How many students in the class do not wear braces?

2. About 2 out of 3 students in a class are going to camp this summer. There are 31 students in the class. How many of these students are going to camp? How many of these students are not going to camp?

3. About 1 out of 2 first graders in our school rides the bus to school. Twenty-nine first graders ride the bus. How many first graders are there *in all*?

4. In a third grade class, 2 out of 4 students bring their lunches to school. There are 28 students in this class. How many students in the class bring their lunches to school? How many do not?

Games People Played

Interview at least 3 adult friends or family members about their favorite games from childhood. Write down each person's name and what you found out about the games.

_____

_____

_____

_____

_____

_____

_____

_____

_____

_____

_____

How does the information you collected compare with our class data about favorite games?

_____

_____

_____

_____

Transportation Data

How One Fourth-Grade Class Gets to School

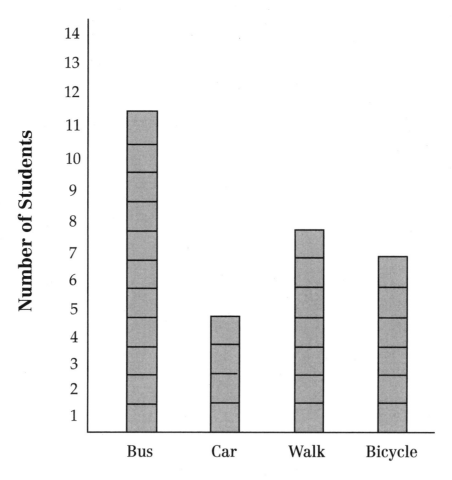

1. Less than $\frac{1}{3}$ of this class takes a bus.   **True**   **False**
   How do you know?   (circle one)

2. More than $\frac{3}{4}$ of the class takes   **True**   **False**
   the bus, rides in a car, or walks.   (circle one)
   How do you know?

Comparing First and Fourth Graders

Choose two of your categories. Write fractions that describe the first grade data and the fourth grade data in these categories. Compare the two fractions.

1. Category: _____

   _____ out of _____ first graders are in this category.

   _____ out of _____ fourth graders are in this category.

Fraction of First Graders	Fraction of Fourth Graders

   Which fraction is bigger? How do you know?

2. Category: _____

   _____ out of _____ first graders are in this category.

   _____ out of _____ fourth graders are in this category.

Fraction of First Graders	Fraction of Fourth Graders

   Which fraction is bigger? How do you know?

What Do You Want to Be When You Grow Up?

Ask one or two adult friends or family members what they would like to be when they grow up.

What did you find out?

_____

_____

_____

_____

_____

_____

_____

_____

_____

_____

_____

_____

_____

_____

_____

_____

_____

_____

_____

1. Working in pairs, print the remaining first and fourth grade data on data cards.

2. Combine all the first and fourth grade data cards. Check to make sure you have the same number of data cards as there are data, or students.

3. Sort all the data according to the categories you already developed. Then redefine categories or make new categories as necessary.

4. Work on the following three activities in any order you would like. Complete all three activities.

   a. Graph the data. Paste your data cards in a permanent display showing the number of first and fourth graders in each category.

   b. Write a few paragraphs about your conclusions. Include any observations you have about the data. Describe how you chose categories. Were any careers hard to categorize?

   c. Choose two categories. Determine the fraction of first graders and the fraction of fourth graders in those categories. Record these fractions on Student Sheet 6, and identify which fraction is bigger in each category. Use cubes, Class Strips, or any other method to help you compare. Explain how you determined which fraction is larger.

CLASS STRIPS

Practice Pages

This optional section provides homework ideas for teachers who want or need to give more homework than is assigned to accompany the activities in this unit. The problems included here provide additional practice in learning about number relationships and in solving computation and number problems. For number units, you may want to use some of these if your students need more work in these areas or if you want to assign daily homework. For other units, you can use these problems so that students can continue to work on developing number and computation sense while they are focusing on other mathematical content in class. We recommend that you introduce activities in class before assigning related problems for homework.

101 to 200 Bingo This game is introduced in the unit *Mathematical Thinking at Grade 4*. If your students are familiar with the game, you can simply send home the directions, game board, Tens Cards, and Numeral Cards so that students can play at home. If your students have not played the game before, introduce it in class and have students play once or twice before sending it home. You might have students do this activity one or two times for homework in this unit.

Ways to Count Money This type of problem is introduced in the unit *Mathematical Thinking at Grade 4*. Here, two problem sheets are provided. You can also make up other problems in this format, using numbers that are appropriate for your students. Students find two ways to solve each problem. They record their solution strategies.

Froggy Races This type of problem is introduced in the unit *Landmarks in the Thousands*. Here, you are provided two problem sheets and one 300 chart, which you can copy for use with the problem sheets. You can also make up other problems in this format, using numbers that are appropriate for your students. On each sheet, students solve the problems and record their solution strategies.

How to Play 101 to 200 Bingo

Materials
- 101 to 200 Bingo Board
- One deck of Numeral Cards
- One deck of Tens Cards
- Colored pencil, crayon, or marker

Players: 2 or 3

How to Play
1. Each player takes a 1 from the Numeral Card deck and keeps this card throughout the game.
2. Shuffle the two decks of cards. Place each deck face down on the table.
3. Players use just one Bingo Board. You will take turns and work together to get a Bingo.
4. To determine a play, draw two Numeral Cards and one Tens Card. Arrange the 1 and the two other numerals to make a number between 100 and 199. Then add or subtract the number on your Tens Card. Circle the resulting number on the 101 to 200 Bingo Board.
5. Wild Cards in the Numeral Card deck can be used as any numeral from 0 through 9. Wild Cards in the Tens Card deck can be used as + or − any multiple of 10 from 10 through 70.
6. Some combinations cannot land on the 101 to 200 Bingo Board at all. Make up your own rules about what to do when this happens. (For example, a player could take another turn, or the Tens Card could be *either* added or subtracted in this instance.)
7. The goal is for the players together to circle five adjacent numbers in a row, in a column, or on a diagonal. Five circled numbers is a Bingo.

101	102	103	104	105	106	107	108	109	110
111	112	113	114	115	116	117	118	119	120
121	122	123	124	125	126	127	128	129	130
131	132	133	134	135	136	137	138	139	140
141	142	143	144	145	146	147	148	149	150
151	152	153	154	155	156	157	158	159	160
161	162	163	164	165	166	167	168	169	170
171	172	173	174	175	176	177	178	179	180
181	182	183	184	185	186	187	188	189	190
191	192	193	194	195	196	197	198	199	200

Practice Page
Three out of Four Like Spaghetti

0	0	1	1
0	0	1	1
2	2	3	3
2	2	3	3

Practice Page
Three out of Four Like Spaghetti

4	4	5	5
4	4	5	5
<u>6</u>	<u>6</u>	7	7
<u>6</u>	<u>6</u>	7	7

8	8	9	9
8	8	9	9
WILD CARD	**WILD CARD**		
WILD CARD	**WILD CARD**		

+10	**+10**	**+10**	**+10**
+20	**+20**	**+20**	**+20**
+30	**+30**	**+30**	**+40**
+40	**+50**	**+50**	**+60**
+70	**WILD CARD**	**WILD CARD**	**WILD CARD**

-10	-10	-10	-10
-20	-20	-20	-20
-30	-30	-30	-40
-40	-50	-50	-60
-70	WILD CARD	WILD CARD	WILD CARD

Practice Page A

Find the total amount of money in two different ways.

    10 quarters
    10 pennies
     5 nickels
    20 dimes

Here is the first way I found the total amount of money:

Here is the second way I found the total amount of money:

Practice Page B

Find the total amount of money in two different
ways.

    2 quarters
39 pennies
15 nickels
  6 dimes

Here is the first way I found the total amount
of money:

Here is the second way I found the total amount
of money:

1	2	3	4	5	6	7	8	9	10
11	12	13	14	15	16	17	18	19	20
21	22	23	24	25	26	27	28	29	30
31	32	33	34	35	36	37	38	39	40
41	42	43	44	45	46	47	48	49	50
51	52	53	54	55	56	57	58	59	60
61	62	63	64	65	66	67	68	69	70
71	72	73	74	75	76	77	78	79	80
81	82	83	84	85	86	87	88	89	90
91	92	93	94	95	96	97	98	99	100
101	102	103	104	105	106	107	108	109	110
111	112	113	114	115	116	117	118	119	120
121	122	123	124	125	126	127	128	129	130
131	132	133	134	135	136	137	138	139	140
141	142	143	144	145	146	147	148	149	150
151	152	153	154	155	156	157	158	159	160
161	162	163	164	165	166	167	168	169	170
171	172	173	174	175	176	177	178	179	180
181	182	183	184	185	186	187	188	189	190
191	192	193	194	195	196	197	198	199	200
201	202	203	204	205	206	207	208	209	210
211	212	213	214	215	216	217	218	219	220
221	222	223	224	225	226	227	228	229	230
231	232	233	234	235	236	237	238	239	240
241	242	243	244	245	246	247	248	249	250
251	252	253	254	255	256	257	258	259	260
261	262	263	264	265	266	267	268	269	270
271	272	273	274	275	276	277	278	279	280
281	282	283	284	285	286	287	288	289	290
291	292	293	294	295	296	297	298	299	300

Practice Page C

1. Two frogs had a race. Minor Frog took 18 jumps of 16. Major Frog took 8 jumps of 36. Who was ahead? How do you know?

2. In a second race, Minor took 9 jumps of 32. Major took 4 jumps of 72. Who was ahead? How do you know?

3. In a last race, Major decided to take jumps of 150. He took 2 jumps of 150. How many more jumps of 150 did he need to reach 300? How do you know?

Practice Page D

1. Three frogs had a race on an imaginary 1000 chart. Opal Frog took 40 jumps of 20. Orange Frog took 32 jumps of 25, and Oiler Frog took 27 jumps of 30. Who was ahead? How do you know?

2. Remember that in the race Opal had already taken 40 jumps of 20. How many more jumps of 20 did Opal need to reach 1000? How do you know?